A FIRST COURSE IN STATISTICS

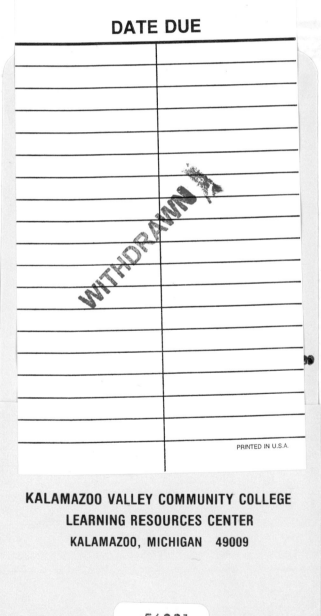

DATE DUE

PRINTED IN U.S.A.

Under the Editorship of Calvin A. Lathan

**KALAMAZOO VALLEY
COMMUNITY COLLEGE**

Presented By

Fenimore Johnson

A FIRST COURSE IN STATISTICS

MARSHALL GORDON

University of North Carolina–Greensboro

NORMAN SCHAUMBERGER

City University of New York at Bronx Community College

MACMILLAN PUBLISHING CO., INC.

New York

COLLIER MACMILLAN PUBLISHERS

London

To O. C. P.

COPYRIGHT © 1978, MARSHALL GORDON AND
NORMAN SCHAUMBERGER

PRINTED IN THE UNITED STATES OF AMERICA

MACMILLAN PUBLISHING CO., INC.
866 Third Avenue, New York, New York 10022

COLLIER MACMILLAN CANADA, LTD.

Library of Congress Cataloging in Publication Data

Gordon, Marshall.
  A first course in statistics.

  Includes index.
  1. Statistics. I. Schaumberger, Norman, joint author. II. Title.
HA29.G684   519.5   77-8891
ISBN 0-02-406680-X

Printing: 1 2 3 4 5 6 7 8   Year: 8 9 0 1 2 3 4

PREFACE

OUR intention was to write a statistics book that was readable, relevant, and suitable to a one-semester course. We want the reader to view statistics as a body of growing useful knowledge that includes techniques to help us better understand aspects of the complex world in which we live. This book was written with the following beliefs in mind:

1. An understanding of basic statistics is essential for every member of our modern society.
2. The social sciences have joined the physical sciences in making considerable use of statistics and statistical techniques.

This book is intended for a general readership. It requires only that the reader understand mathematics at a level of first-year high school algebra.

We would like to thank Tamara Bruce for her patient typing of the manuscript.

We are grateful to the Literary Executor of the late Sir Ronald A. Fisher, F.R.S., to Dr. Frank Yates, F.R.S. and to Longman Group Ltd., London, for permission to reprint Table III from their book *Statistical Tables for Biological, Agricultural and Medical Research* (6th edition, 1974).

M. G.
N. S.

CONTENTS

Introduction

In responding to the wide range of uses of elementary statistics in society and the social sciences, we have chosen to eliminate in this book some of the more highly complex statistical techniques and instead have included chapters on survey taking and nonparametric statistical tests in the hope that the reader will undertake a survey as a way of developing a personal understanding of statistics.

The book can be viewed as containing three sections. The first section, Chapters 1 to 3, discusses statistics as a necessary way of understanding a world that is more and more number-oriented. We begin by organizing information graphically. These pictures of numerical information, or graphs, provide us with a way to organize data meaningfully. Some of these graphs, such as the bar graph, you have probably seen before. Others may be new to you. What is important in this discussion is the construction of these graphs. You will see that converting collected information into meaningful information is really both an art and a science. After the information has been presented graphically, we shall examine it arithmetically. That is, we begin to ask if we can find one or two numerical values that will provide us with another way of understanding the information. You will see how information can be reorganized to help us understand how a particular score can be compared to a large collection of scores, and how a collection of scores can be reorganized to provide greater insight into the situation.

Whereas the first section emphasized ways of simplifying and describing information, the second section, Chapters 4 to 6, emphasizes ways in which we can make predictions based upon the information at hand. We shall consider probability, based on experience and logic, because this area of mathematics is essential to a study of statistics and marks a major effort by human beings to organize information about events in terms of the chance of these events occurring or not. Chapters 5 and 6 examine two important theoretical distributions: the binomial and normal distributions. These distributions are very valuable in helping us test our beliefs and hypotheses about certain statistical and probabilistic situations. Additionally, when we do not have enough information to construct a hypothesis to be tested, we can use these distributions to determine numerical estimates of what might be true of a population given the little information we have from a sample of the population.

1

In the third section, Chapters 7 and 8, we examine other statistical tests that can be used in evaluating our hypotheses and look at some of the important aspects of taking a survey. The statistical tests in this section are labeled "distribution-free" tests. That is, in contrast to the binomial and normal distribution tests, which require that the populations from which we collect our data satisfy certain requirements, the distribution-free tests are not as stringent and thus have wider appeal. Although the chapter on survey taking is left for last, it is mentioned very early in the reading so that if you have any intention of taking a survey, certain sections should be read at your earliest convenience. A good deal of effort is required in constructing an acceptable questionnaire and in collecting the data. If you decide to take a survey, all of the probability and statistics in this book may be of value. We have found that when we take a survey we find out a little more about what is going on around us and thus learn a little more about ourselves.

The chapters include many illustrative problems, which may be found at the end of sections and are numbered within the section; for instance, the first problem in Section 3.3 is numbered 3.3.1. The figures and tables are also numbered by section. Most of the chapters conclude with a chapter summary, a formula review, and a problem set that should serve as a way for you to reflect on the important considerations in each chapter.

Chapter 1
A Changing World

1.1 The Certainty of Chance

One thing we can be certain of is that we take chances. The chance of success usually depends upon our understanding of the situation in which we find ourselves. The greater our understanding, the more likely it is that we shall succeed. This applies to us both as individuals and as citizens of the changing world in which we live. Governments all over the world adapt their economic, social, and political relationships to other governments as new information is collected and new directions are suggested. Similarly, we are constantly learning, and thus altering, our understanding. The way we approach a problem today may differ from the way in which we would deal with the same problem tomorrow. The changing world is a continual reminder that new information may require changing our decisions and taking chances.

When we are faced with a situation that requires a decision to be made, many choices must be considered. For example, if we have decided to continue our education, we must decide how our career goals are likely to develop and what training would be most useful and rewarding. If we decide in favor of further schooling, we must look at our financial situation in light of what the school expenses will be. Depending upon our situation, the plan of action is probably among the following options:

1. School full-time.
2. School full-time, working part-time.
3. School and work part-time.
4. School part-time, working full-time.
5. Work full-time and save for full-time school in the near future.

Each one of these options requires additional considerations. We should also consider our ability, the school's location, offerings, social atmosphere, and many other characteristics.

All the information we collect represents the data we could use in making our decision. Some numerical data we might consider important are the distances between a large metropolitan area and the possible schools we are considering—especially if transportation is a major concern. For example, suppose that five schools were being considered, whose distances to a metropolitan area were (in miles): 7, 12, 3, 8, and 15. We might want to limit the distance we would have to travel to 10 miles. In this case we would narrow the variability; that is, the size of the variation in

3

distance would be limited to the three schools that are 3, 7, and 8 miles away, respectively.

Whether we have to make a decision as difficult as choosing a school or as simple as choosing what to wear, we must make use of the data we have. Very often, we reach decisions without having all the information. And sometimes the information that is available is confusing. For example, a store selling stereo equipment suggests that we should buy from them because "The 100 managers of our stores have a total of over 2,000 years of experience!" To make decisions wisely and increase our chance of success, we need to collect all the available helpful information, that is, the relevant data, and then organize these data into a clear and comprehensive picture of the situation. *Statistics* as a body of knowledge is a science that has developed techniques to collect and organize data and ways to make decisions in the face of uncertainty. Statistics may also mean numerical data—our statistics for the distance of the school to the metropolitan area was 7, 12, 3, 8, and 15 miles.

1.2 Samples from Populations

Our decisions are usually based on only a *sample* of all the possible data, that is, a part of all the possible information. Sometimes we make judgments on too little information, such as guessing how a baseball player will do for the season based on his first 10 times at bat. A better understanding of the baseball player's ability would be gained by looking at what he did every tenth time he came to bat during the season. If we collected these data, not only would we have more information to make a judgment than if we considered only the first 10 times at bat, but the data would better reflect how the batter did for the entire season.

If the sample data we collect are representative of the total data—the *population* data—then statistical methods can be used to evaluate the situation based on only a portion, or sample, of the information, and the conclusions drawn can give us insights into the total situation. For instance, professional pollsters can take a sample of about 1,550 voting-age people from the total population of approximately 80,000,000 voters and predict with a high degree of accuracy how the total voting population will vote.

In the study of statistics, "population" does not always refer to a population of people. A biologist may be interested in studying the reaction of white blood cells to a certain chemical. The population in this case would be the cells that were injected. And the biologist would probably choose a sample of these cells to examine the chemical effects. An efficiency engineer may be interested in checking special machine parts to determine if they are of good quality. The engineer could choose a sample of these parts for examination from the entire population of parts produced during a specific period of time. We could formally define a population and a sample as:

Population: a collection of all objects sharing some common characteristic(s).

Sample: a collection of objects taken from a population.

When we make decisions, we count on common sense, based on experience, to help us. However, we must remember that "common sense" told us that the earth was the center of the universe, all women belong in the home, and punishment makes students learn. Greater awareness has led us to dismiss all three ideas as being poorly conceived. For one, after collecting sample data on student achievement and using statistical techniques to analyze the data, we now believe that success (reward) is a more effective way to promote learning. We shall, during the course of this book, develop statistical techniques to help us examine and draw conclusions from experience rather than depend solely on common sense.

1.3 Descriptive Statistics and Inferential Statistics

The study of statistics encompasses two broad areas:

1. *Descriptive statistics*—This area of statistics is concerned with ways of collecting and organizing data. We can represent the data in graphic form and determine the location, shape, and spread of the data. Judgments are not made about the population from which the sample data are taken. Rather, we simply describe the data. For example, we collect the grades of the students in a class over the course of 1 year. Then we could describe how the grades are distributed, the central tendency of the grades, and the spread of the grades.

Suppose that everyone in your class took a reading of their pulse when they got up in the morning. The population of pulse beats would vary. People who are in excellent shape usually have low pulse beats; for example, some soccer players and astronauts have pulse beats of between 30 and 40 beats per minute. The "average" person has around 72. If the students in the class jogged for 1 minute and then took their pulse reading again, the readings would be different from the at-rest readings due to the physical exertion. If the two collections of data were each organized in a *frequency distribution*, which is a record of the number of times something happens that satisfies some condition (e.g., four people had a rest pulse of 72), a comparison could be made of the two distribution shapes, central tendencies of each distribution, and the location of the data in each distribution.

2. *Inferential statistics*—This area of statistics is concerned with making inferences; that is, making judgments or predictions about the populations from which the representative samples were taken. For example, in

examining the grades of a student received in the first year (a sample), a prediction can be made as to the student's grade-point average at the conclusion of his/her studies. Since this is a prediction based upon only a portion of the population data (all his/her grades), there is a degree of uncertainty in the prediction. Another instance of making inferences based upon an examination of a particular sample occurs in the testing of new drugs by a pharmaceutical company. As long as the sample is representative of the population, the results of the sample's reaction to the drug can be used to make judgments regarding the entire populations' reaction to the drug. However, if the sample is not representative, widespread use of the drug could produce harmful effects that were not observed in the analysis of the sample data.

The crucial elements in inferential statistics are choosing sampling procedures, selecting the correct test to analyze the data, and measuring the error inherent in making judgments based on sample data. If the sampling plan contains the property of *randomness*, we can make judgments about the population and feel that our findings reflect the population. We can define a random sample as:

Random Sample: a sample in which every member of the population has an equal chance of being chosen.

A random sample does not imply a haphazard sample. For instance, standing on a corner and collecting opinions is a haphazard way of finding out how a community feels about a pending bill in the legislature. The sample is not random or representative of the community; rather it is a *biased* sample inasmuch as it is a sample of only those people who happen to pass by this corner at this particular time. Any judgment that is based on data from a biased sample will contain a large chance of being in error. If a random sample found that 35 per cent of the community was in favor of the pending bill, we could estimate that about 35 per cent of the total population of the community would be in favor of the bill. Whereas if the sample had been a biased one, the 35 per cent findings could be totally misleading as to the true feelings of the total voting population. The interested reader should read Section 8.2 at this time for a discussion of sampling techniques.

Usually, we can only examine a sample from a population, since the time and money needed to examine the entire population is too great, especially if the population is large. The area of inferential statistics provides statistical techniques to help us analyze and make judgments about the characteristics of a large body of information from an examination of a well-chosen (random) sample.

1.4 Problem Set

1.4.1 From looking at television, reading newspapers, and listening to people, can you find any statements that are based on biased samples? What about forming opinions concerning a person's character, intelligence, or ability based upon a single statement or action?

1.4.2 Everyone in your class should determine his/her pulse rate after rising in the morning. The data should be organized into a frequency distribution; for example, three people had 64, four people had 67, and so on.

1.4.3 Create a separate frequency distribution for the students in the class of their pulse rates found after jogging for 30 seconds.

1.4.4 Take the rest pulse data and separate them into categories of male and female; smokers versus nonsmokers, and so on. Can you detect any differences between the groups?

1.4.5 Perform, for the jogging pulse data, the analysis suggested in Problem 1.4.4. Can you detect any differences between the groups?

Chapter 2

Data into Meaningful Information

2.1 Collecting Data

In the process of thinking, we change our random observations into meaningful information; but the very process of collecting and organizing information has its difficulties. At times, because of the magnitude of the situation, important information is overlooked. For instance, an international airport was built in Tokyo, Japan, in 1972 but, as late as mid-1974, a single plane had not landed or taken off from the airport. Someone had forgotten to include gas lines to the airport, so no gasoline was available for the planes. The owners of the airport continue to pay their creditors $80,000 daily in interest as a reminder of their mistake. This situation is an extreme one, but are there not some times when we all overlook information because we did not organize the information carefully enough?

Aside from not overlooking important information, we must raise certain questions when organizing information. Is the information clear? Have the data been collected without building in any bias? Is the information representative of the group under consideration? What is the size of the group the data came from? Vague facts can be found in many places. For example, one type of television commercial states: "8 out of 10 doctors surveyed say—." We can ask: "Are all these doctors trained in this area? How did they choose the sample—were they chosen from all doctors in the country? State? Were there only 10? More generally, did they send out a questionnaire to a random sample of doctors?"

The question of how many data to collect before making a decision also requires careful consideration. The national opinion poll surveys today ask about 1,500 voters who they think will be elected president and the findings are very close to the actual population vote. On the other hand, in 1936, the *Literary Digest*, after polling over 2,000,000 voters, predicted that Alfred Landon would beat Franklin Roosevelt. How could this happen in view of the fact that the *Literary Digest* polled approximately .029 (2,000,000/70,000,000) of the voting population and today's pollsters poll only about .00002 (1,500/80,000,000) of the voting population? The answer can be found in the fact that the *Literary Digest*'s sample did not

8

reflect all the voters—just the wealthy portion of the society. This biased sampling approach could indeed have contributed to the results! For, upon hearing the results of the poll, Landon might have campaigned less and Roosevelt might have compaigned more. There is growing awareness of the possible effects of poll taking on people's attitudes. For example, the results of a poll taken to determine who the French people favored for president after Pompidou's death in 1974 was ordered not to be released until after the election so as not to influence the voters. The fact that many people are influenced by the results of polls is an important consideration, especially if the sample is biased. While the size of the sample is important, the methods of sampling are at least as important. (The reader interested in taking a survey should read Sections 8.1 and 8.3 at this time.)

Purchasing is one area in which we carry out an informal level of sampling to obtain needed information. For instance, if you planned to buy a used car, to whom would you go for advice? What would constitute your sample data? If you asked your friends, is the sample biased? If you only look at cars in your immediate area, is the sample biased? If you examined technical and consumer reports, are these data relevant? Would the opinions of automobile repair shops be good? Suppose that you wanted to buy a particular car: would the opinion of used car dealers who do not sell the car you are interested in be valuable? What about used car dealers who do sell the car you want? Clearly, the problem of collecting the relevant data is an involved task. Moreover, once you have collected all the data, they should be organized to help you make wise decisions.

2.2 Organizing Data—Bar Graphs

One of the basic methods for organizing numerical data and representing them in pictorial form is the *bar graph*.

The following example shows how organizing data can help us to understand complex situations. Suppose that an employer agreed to provide his employees with a life insurance policy. Three plans were available:
1. Every active employee gets a $2,000 policy.
2. Every active employee who is at least 24 years of age gets a $5,000 policy.
3. Every active employee who is at least 30 years of age gets an $8,000 policy.

Plan 1 would cost the employer $5 per policy; plan 2, $11; and plan 3, $19.

It is obvious that in order to make a wise decision we need to know the ages of the employees. Table 2.2.1 shows the ages of all employees, taken from an alphabetical listing. The table presents the population data; that is, all the employee ages are listed. All the ages have been rounded to the nearest year.

Table 2.2.1. Ages of 100 Employees

18	20	33	17	23	34	16	16	22	19	34
34	36	43	17	19	16	28	16	27	24	31
37	42	16	18	39	19	21	21	19	22	25
27	43	25	38	17	20	18	23	16	30	17
27	19	35	40	20	18	23	22	18	29	32
36	17	21	28	16	36	40	22	20	24	18
32	23	35	17	23	16	27	18	30	17	22
20	25	32	19	24	24	29	18	33	20	21
21	25	33	29	21	23	28	20	38	24	23
22										

Without organizing the data in some meaningful way, we cannot readily make any judgment regarding the age distribution. One way to order the data is to count the number of occurrences, that is, the frequency, for each age, and arrange the scores in a table known as an *ungrouped frequency distribution*, as in Table 2.2.2.

Table 2.2.2. Ungrouped Frequency Distribution of Ages of 100 Employees

Age	Number of employees	Age	Number of employees	Age	Number of employees
16	8	25	4	34	3
17	7	26	0	35	2
18	8	27	4	36	3
19	6	28	3	37	1
20	7	29	3	38	2
21	6	30	2	39	1
22	6	31	1	40	2
23	7	32	3	41	0
24	5	33	3	42	1
				43	2
					100

Now we can determine the cost of each plan. Plan 1 would cost $5 × 100 employees = $500. From the frequency distribution, we observe that there are 45 employees who are at least 24 years of age (24 years or older), and so plan 2 would cost $11 × 45 = $495; and since there are 26 employees who are at least 30 years of age, plan 3 would cost $19 × 26 = $494. If you were the employer, which plan would you choose? If you were an employee and you were 17 years old, which plan would you choose? If you were 23 years old? If you were 40 years old?

Let us assume that the employer would like to make as many of his employees happy as he can. Hence he might choose plan 1, since this would give all 100 employees a policy. However, some of the older employees might complain that they would be receiving the same benefits as the younger employees. Thus the employer might consider choosing a new plan, plan 4, by which employees 16 to 19 years old receive $3,000; 24 to 27, $4,000; 28 to 31, $5,000; 32 to 35, $6,000; 36 to 39, $7,000; and 40 to 43 $8,000.

In order to determine how many employees are in each group, we can either create a grouped frequency distribution or a bar graph. We shall use a bar graph, where the height of each bar represents the frequency of each group (Figure 2.2.1). A bar graph, if it is to be informative, should contain between 5 and 20 categories. Fewer than 5 categories tends to cause too much lumping of data, while more than 20 creates too many categories to work with efficiently.

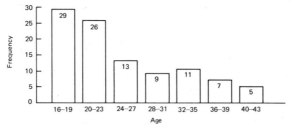

Figure 2.2.1. *Bar Graph of 100 Employees by Age*

Under plan 4, the 16 to 19 age group costs $3 for each member; the 20 to 23 group, $4 for each member; 24 to 27 group, $5 each; 28 to 31 group, $6 each; 32 to 35 group, $8 each; 36 to 39 group, $9 each; and 40 to 43 group, $10 each. The total cost is found to be: $29 \times 3 + 26 \times 4 + 13 \times 5 + 9 \times 6 + 11 \times 8 + 7 \times 9 + 5 \times 10 = \$511$. Although the plan costs a little more, the employer might feel that the increase in worker morale outweighs the extra cost and that it is a better plan than any of the others.

A *grouped frequency distribution* can be modified to show the proportion or relative frequency of employees in each age group. As such, it is known as a *grouped relative frequency distribution* (Table 2.2.3). We can think of each bar in the bar graph as containing an area equal to the relative frequency of each age group. For example, 11 per cent of the total area (100 per cent = 1) is contained in the 32 to 35 bar as there are 11 workers in that age group, or relative to the total (100) there is 11/100 of the total group. Notice that, when we add up all the relative frequencies, the total is 1. This should be the case, since 100 per cent $= \frac{100}{100} = 1$ (all) of the cases have been accounted for. We find from the relative frequency table that 23 per cent of the workers $(.11 + .07 + .05)$ have insurance benefits of at least $6,000; that is, every worker who is 32 years of age or older has these benefits.

Table 2.2.3. Grouped Relative Frequency Distribution of 100 Employees' Ages

Age group	Frequency	Relative frequency
16–19	29	$\frac{29}{100} = .29$
20–23	26	$\frac{26}{100} = .26$
24–27	13	$\frac{13}{100} = .13$
28–31	9	$\frac{9}{100} = .09$
32–35	11	$\frac{11}{100} = .11$
36–39	7	$\frac{7}{100} = .07$
40–43	5	$\frac{5}{100} = .05$
	100	$\frac{100}{100} = 1.00$

Both ungrouped and grouped frequency distributions along with bar graphs demonstrate how raw data can be organized into meaningful information. The following discussion demonstrates the advantages and disadvantages of the bar graph.

EXAMINING THE BAR GRAPH

1. What per cent of the group is between 20 and 31 years of age (inclusive)? Since there is a frequency of 26 in the 20 to 23 bar, 13 in the 24 to 27 bar and 9 in the 28 to 31 bar, there are $26 + 13 + 9 = 48$ people in this age group. Thus $\frac{48}{100} = 48$ per cent of the workers are in this group.

2. What proportion of the workers are less than 40 years of age? We can approach this question in two ways: (1) we could add up all the relative frequencies for every age group except the last $(.29 + .26 + \cdots + .07)$, or (2) we could see that since the total is 1—we would just subtract the proportion in the 40 to 43 age group $(.05)$ from the total; that is, $1 - .05 = .95$. Everyone (100) minus the number in the 40 to 43 age group (5) is less than 40 years of age; that is, $100 - 5 = 95$ of the 100 people, $\frac{95}{100} = 95$ per cent, are less than 40 years of age.

3. How many workers are at least 39 years of age? We cannot determine this number from the bar graph, since there is no bar with 39 as its smallest value. That is, 39 falls in the 36 to 39 group and the frequency is the grouped frequency of all workers from 36 up to and including the 39-year-old workers. To find the answer, we must look at the individual frequency distribution (Table 2.2.2): $1 + 2 + 0 + 1 + 2 = 6$. This situation illustrates the problem of losing sight of the individual in the group, something that is not confined to the study of statistics.

In our grouped frequency distribution, we created seven *class intervals*: 16–19, 20–23, 24–27, 28–31, 32–35, 36–39, and 40–43. Each class interval has a *class width* = 4; that is, each interval contains four years. For example, the first class interval contains the years 16, 17, 18, and 19.

PROBLEMS

2.2.1 Earlier we said that between 5 and 20 classes should be created graphically. In our example, we chose seven. To determine the possible class widths, we first find the *range* of scores; that is, the difference between the largest and smallest score—in our example, $43 - 16 = 27$ is the range. Then we divide by the largest and smallest number of acceptable classes—for our example, any class width between $\frac{27}{20} = 1.35$ and $\frac{27}{5} = 5.4$ is acceptable. We chose 4 as the class width.

Use the data in Table 2.2.2 and a class width of 2 units (years) to determine:
(a) A grouped frequency distribution.
(b) A relative frequency column.
(c) The per cent of workers who are at least 18 years of age; 30 years of age.
(d) A *cumulative relative frequency* column; that is, next to each relative frequency put the value of the cumulative (accumulated) relative frequency value that is the sum of the relative frequencies up to and including the relative frequency on the same line. For example, the first two lines of your distribution should read:

Age group	Frequency	Relative frequency	Cumulative relative frequency
16–17	15	.15	.15
18–19	14	.14	.29 (.15 + .14)

(e) What per cent of the workers are at most 35 years of age? What per cent are over 35 years of age?

2.2.2 *Class Project*: Write down the time, to the nearest hour, you read this sentence. For example, if it is 2:27 P.M., it becomes 14:27—and to the nearest hour it is 14. Create a class grouped frequency distribution and see if any conclusions can be made regarding study habits.

2.2.3 *Class Project*: Take your rest pulse data and jogging pulse data and create two grouped frequency distributions, and see if any conclusions can be suggested.

2.3 Organizing Data—Histograms

In the life insurance problem just considered, the ages were rounded off to the nearest year. That is, people who were 19 years and less than 6 months were put in the 19-year-old group, while those who were 19 years old and more than 6 months were put in the 20-year-old group. When we consider variables such as time, weight, or length, we are talking about *continuous variables,* variables that can have any value over an interval. Contrast these with *discrete variables,* variables that have a countable number of values. Examples of discrete variables are: the number of members of a family or the number of students in a class. For any situation where a discrete variable is considered, we can use a bar graph to depict the distribution. Histograms can be used when the description is of a continuous variable. A *histogram* is a bar graph with adjacent bars sharing the same value. We can convert a bar graph to a histogram by adjusting the class intervals.

In our insurance problem, the class width is 4. We can keep this width and by making the smallest age, 16, $\frac{1}{2}$ unit less, or 15.5, and the largest age 43.5, $\frac{1}{2}$ unit more, the histogram appears as shown in Figure 2.3.1. The class frequencies are the same as they were in the bar graph, Figure 2.2.1.

Figure 2.3.1. *Histogram of Ages of 100 Employees*

When describing a continuous variable in the form of a histogram, we can create class intervals so that no scores fall on the boundaries of any interval. This was the case in Figure 2.3.1. Thus, to create a histogram, we:
1. Determine the lowest and highest scores.
2. Create a new lowest and highest score by making the lowest score $\frac{1}{2}$ unit less and the highest score $\frac{1}{2}$ unit more.
3. Determine the range using the new scores (new highest score minus new lowest score).
4. Find the acceptable class widths by dividing the range by 5 and 20 (the minimum and maximum number of class intervals).
5. Choose an acceptable class width.
6. Starting with the new lowest score, add the class width, creating class intervals until the new highest score is obtained.

This approach is used in examining the following situation. Twenty-five students were asked to draw a line segment that each student thought was about 30 centimeters long. The students knew that there are approximately 2.54 centimeters per inch. Their responses were measured in centimeters and given in Table 2.3.1. (Notice that while an inch is broken into 8 or 16 parts on a regular ruler, each centimeter is broken into 10 parts, which makes it more convenient to use.)

Table 2.3.1. Estimates of 30 Centimeters by 25 Students

26.4	29.1	35.2	23.7	28.2	34.1	32.7	26.8	30.9	27.5	30.4	26.6	28.9
33.4	37.0	25.6	27.3	22.7	32.8	34.3	31.8	27.7	29.4	25.1	32.0	

Because none of the scores were repeated, a grouped frequency distribution would be of more help in describing the data than an ungrouped distribution. This would also be true if only a few of the scores were repeated. The lowest score is 22.7 and the highest score is 37.0. Thus the new low and high scores, created to eliminate scores falling on two intervals, are 22.65 and 37.05, respectively. The new range is $37.05 - 22.65 = 14.40$. (Notice that the range of the raw data is 14.3.) The acceptable class widths are any choices between $14.40/20 = .72$ and $14.40/5 = 2.88$. If we want to have eight intervals, the class width is $14.40/8 = 1.80$. Starting with 22.65 and adding 1.80, we can create a frequency distribution, as in Table 2.3.2. We have included the *class marks*, which are the midpoints of each interval, and the relative and cumulative frequency columns. Each class mark (midpoint) is found by adding one half the width of the interval to the lower bound of each interval. For example, $22.65 + \frac{1}{2}(1.80) = 22.65 + .90 = 23.55$ is the first class mark. Each additional class mark is found by adding

Table 2.3.2. Grouped Frequency Distribution of 25 Estimates

Class intervals	Mark (midpoint)	Frequency	Relative frequency	Cumulative relative frequency
22.65–24.45	23.55	2	$\frac{2}{25} = .08$	.08
24.45–26.25	25.35	2	$\frac{2}{25} = .08$	.16
26.25–28.05	27.15	6	$\frac{6}{25} = .24$	.40
28.05–29.85	28.95	4	$\frac{4}{25} = .16$	.56
29.85–31.65	30.75	2	$\frac{2}{25} = .08$	.64
31.65–33.45	32.55	5	$\frac{5}{25} = .20$	.84
33.45–35.25	34.35	3	$\frac{3}{25} = .12$	.96
35.25–37.05	36.15	1	$\frac{1}{25} = .04$	1.00
		25	1	

the class width to the preceding class mark. For example, $23.55 + 1.80 = 25.35$ is the second class mark. A relative frequency histogram using class marks in place of the class boundaries depicts the relative frequency distribution of the grouped data as shown in Figure 2.3.2.

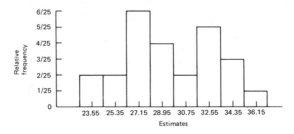

Figure 2.3.2. *Relative Frequency Histogram*

The cumulative relative frequency column provides us with additional information. For example, 64 per cent of the guesses were 31.65 or less. A graphical representation of the cumulative relative frequency column is known as a relative frequency *ogive* curve. The ogive curve is created by sketching a curve through the endpoints of each interval at a height equal to the cumulative relative frequency (Figure 2.3.3). We can approximate that 70 per cent of the guesses were 32 and below.

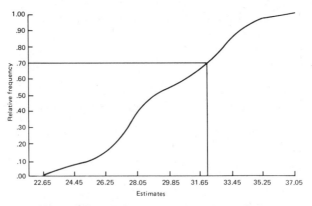

Figure 2.3.3. *Relative Frequency Ogive Curve*

The frequency distribution and the ogive curve can also be used to determine the answers to the following questions.

EXAMINING THE RELATIVE FREQUENCY HISTOGRAM

1. What percent of the guesses were at most 24.4 centimeters? All the scores that fell to the left of 24.45 would be at most 24.4. Since there are two scores, this is 8 per cent of the scores.

2. What per cent of the guesses fell between 28 and 31.6 centimeters? The area under the histogram from 28.05 to 31.65 is found to be $\frac{4}{25} + \frac{2}{25} = \frac{6}{25} = 24$ per cent of the total area.

3. What proportion of the scores were at least 24.4? We could count the number of scores from 24.4 to 37.0 and divide by 25—or, since the total area under the relative frequency histogram is 1, and we know from question 1 that 8 per cent of the area is in the region from 22.65 to 24.4, we can find that $1 - .08 = .92 = \frac{23}{25}$ of the scores were at least 24.4.

4. How many scores were at least 28? We cannot answer this question by examining the histogram because 28 is not an endpoint of an interval. From the ogive, we find that about 40% of the scores fall below 28, so that 60 per cent of the scores are above 28. In Chapter 3 we shall determine a method for examining histograms which will provide a good approximation for finding specific values that are not class boundaries.

PROBLEMS

2.3.1 A student wanted to get an idea of how he spent his money over the course of a month. The following 30 values are the amounts, in dollars, of the money spent in September: 6.54, 8.10, 12.14, 9.00, 4.30, 3.51, 5.00, 12.86, 3.80, 6.75, 7.55, 2.90, 5.80, 10.20, 11.55, 3.40, 2.83, 9.70, 8.15, 4.55, 6.51, 8.25, 3.90, 4.27, 5.80, 9.65, 7.23, 6.40, 3.25, and 9.80.
  (a) Is this variable discrete or continuous?
  (b) Would it be helpful to create an ungrouped frequency distribution? Why?
  (c) Create a frequency histogram with a class width of 1.

2.3.2 Round off the amounts spent to the nearest unit (for example, 2.83 becomes 3). Using the same class length as in Problem 2.3.1(c), create a grouped frequency distribution. Is there any real difference between the distributions in Problem 2.3.1(c) and this example?

2.3.3 Using Problem 2.3.1(c), determine:
  (a) The relative frequency column.
  (b) The cumulative relative frequency column.
  (c) The number of days that more than $6.82 was spent; the proportion of time that more than $3.82 and less than $7.82 was spent; and the number of days that more than $11.82 was spent. What per cent of the area is located in the first four class intervals?

2.3.4 Use the rest pulse data you collected and create:
  (a) A frequency histogram with a class width of 4.
  (b) A relative frequency column.
  (c) A cumulative relative frequency column.

2.3.5 Use the jogging pulse data and reconsider Problem 2.3.4.

2.3.6 Can you draw any conclusions in comparing your findings in Problems 2.3.4 and 2.3.5?

2.3.7 Would you choose a different class width? Why?

2.4 Special Types of Distributions

Certain easily recognizable shapes are frequently taken by histograms when they are used to examine real-world situations. We shall now look at these more common distributions and discuss their shapes. In future chapters we shall reexamine some of these distributions in terms of their specific location and spread, as they are valuable distributions in the study of statistics.

RECTANGULAR (UNIFORM) DISTRIBUTION

A uniform (equal) distribution of scores creates a histogram that has the shape of a rectangle. The frequencies or relative frequencies of each category are the same. For example, if a product containing parts *A*, *B*, and *C* were all produced by machine, each day's production should contain an equal number of each of these parts (Figure 2.4.1). Can you think of any other situations where the distribution will be uniform?

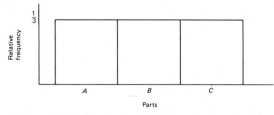

Figure 2.4.1. *Example of Rectangular Distribution*

NORMAL DISTRIBUTION

If you were the manager of a shoe store and you could choose from 20 available styles, would you order the same number of each style? Would you order the same number of sizes for each style (uniform distribution)? Whereas the first question requires knowing the likes and dislikes of your customers, the second question can be answered given a large-enough sample from the population. The answer is no, inasmuch as shoe sizes— along with many other measurements, such as height and IQ—form a pattern that approximates a normal distribution. In Table 2.4.1 we have a distribution of scores that approximates a normal distribution pattern.

One of the essential characteristics of the normal distribution curve is its symmetry. If we draw a vertical line through the center and fold along the line, the two halves would match exactly. This comes from the fact that scores equidistant from the center of the distribution have the same frequency or relative frequency. In Table 2.4.1 the center of the distribution is the class mark of 16. The class marks 2 units from the center $(16-2=14$ and $16+2=18)$ have the same frequency (9): the class marks 4 units from the center (12 and 20) have the same frequency; and so on.

Table 2.4.1. Example of Normal Distribution

Class mark	Frequency
10	2
12	5
14	9
16	12
18	9
20	5
22	2

A graph of the distribution is shown in Figure 2.4.2. Notice that in this histogram its greatest frequency, 12, is at the center and the distribution "tails off" in both directions from the center. These are the other properties of a normal distribution. More will be said later about this important distribution. Can you think of other situations where the distribution is symmetrical, has its greatest frequency in the center, and tails off as we move away from the center?

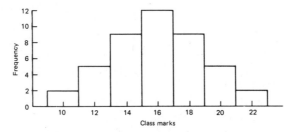

Figure 2.4.2. *Histogram for a Normal Distribution*

J-CURVE DISTRIBUTION

A J-curve distribution is, as one would expect, a distribution that has a shape similar to a J. This distribution is characterized by the fact that the minimum frequencies occur at the left, and as the distribution scores

Table 2.4.2. J-Curve Frequency
Distribution

Class mark	Frequency
55	1
65	2
75	7
85	10
95	16

increase, the frequencies increase, so that the maximum frequencies occur at the right. For example, if a class, in general, scored high on an exam, the distribution would approach a J-curve (Table 2.4.2 and Figure 2.4.3). Other distributions that would form a J-curve shape are the average daily wage of migrant farm workers over the last 50 years, and the number of books published over the last 100 years in 5-year intervals.

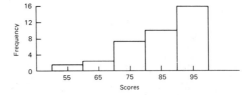

Figure 2.4.3. *J-Curve Frequency Distribution*

A reverse J-shape (Ɩ) distribution is one in which the maximum frequency occurs at the left side of the distribution (smaller scores) and then tails off so that the minimum frequencies occur at the right side (higher scores). As would be expected, this distribution is the reverse of the J-curve. For example, the number of people who contracted polio beginning with the year the Salk vaccine was marketed decreases as the vaccine becomes more and more available. Other examples include oil consumption over a period of years in which a cheaper form of energy is being perfected and being made available; the number of new births of a species that is becoming extinct. Can you think of other examples that have J-curve and reverse J-curve distributions?

SKEWED DISTRIBUTION

Recall that the normal distribution shape is symmetrical; that is, we can divide the distribution into two equal-shaped halves by drawing a line vertically through the center of the distribution. A skewed distribution is one in which one side of the distribution tails off. If it tails off to the right, the distribution is referred to as being "skewed to the right," and if it tails off to the left, it is referred to as being "skewed to the left." For example, Figure 2.4.4 is the shape of a distribution skewed to the left, and Figure 2.4.5 is the shape of a distribution skewed to the right. When we draw a vertical line through the midpoint of the range, one side of a skewed distribution would contain much less of the total area than the other. The distribution of scores of a class on a difficult test might appear as in Figure 2.4.4, while the distribution of scores on an easy test might appear as in Figure 2.4.5.

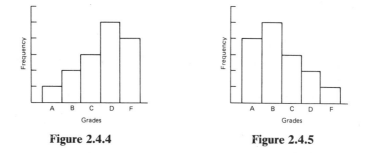

| **Figure 2.4.4** | **Figure 2.4.5** |

2.5 Pictures Are Worth a Thousand Words

Presentation of data can take many forms. Depending upon who is presenting the information and why, information can be "pictured" in such a way as to imply a particular point of view, which may or may not be the case. Some examples of this situation follows.

GRAPHS WITH "BLINKERS"

When the scope of the information presented is narrow with respect to time or circumstances, the inference we draw might be quite inappropriate. The following bar graphs describe the financial situation of companies *A* and *B* (Figures 2.5.1 and 2.5.2). Which company looks better to you?

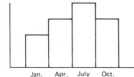

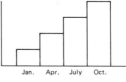

Figure 2.5.1. *Frequency Distribution of Earnings in 1974 for Company A*

Figure 2.5.2. *Frequency Distribution of Earnings in 1974 for Company B*

From this comparison, company *A* would seem to be doing better than company *B*. However, the additional information that company *B* is just beginning and that company *A*'s earnings had been down for the three previous years, might cause one to take a second look. When looking at a bar graph, we must always ask: "Is too little information presented to make a reasonable judgment?"

SHRINKING AND STRETCHING

Sometimes when we look at data presented graphically, our eyes make judgments before our brain evaluates the situation. Which of the two situations represented graphically in Figures 2.5.3 and 2.5.4 appears to be more dramatic? Why?

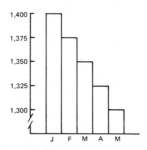

Figure 2.5.3. *Frequency Distribution of Sales*

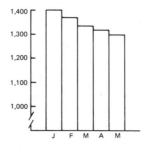

Figure 2.5.4. *Frequency Distribution of Sales*

The two figures present the same information—could you tell that from just an "eye-view"? The vertical axis was stretched in Figure 2.5.3 and shrunk in Figure 2.5.4. By the appearance, Figure 2.5.3 shows a significant decline, while Figure 2.5.4 shows only a slight decline. The horizontal axis can also be manipulated to produce certain results. Create two bar graphs and, by stretching and shrinking the horizontal axis, show that two different impressions can be made.

GROWTH VERSUS RATE OF GROWTH

Although it may sound strange, there are situations that at a glance are different from the way they appear on closer examination. Such a situation occurs when growth increases but the rate of growth decreases. An example of this can be seen in terms of airplane deaths in the United States at 10-year intervals from 1920 to 1960 (excluding deaths due to war), where the number of deaths can be seen to be increasing. However, since the number of passenger miles has increased so greatly, the rate of passenger deaths per 100,000 passenger miles has, in fact, decreased considerably. When we are presented with data that describe a certain growth, we must also consider the growth rate.

2.6 Summary

In this chapter we explored ways of graphically presenting information. We used tables, bar graphs, and histograms to aid us in understanding and interpreting raw data. We looked at some distributions that occur frequently in the real world as well as how graphs can be used to present a particular viewpoint.

A bar graph can be useful to present a frequency distribution of a discrete variable. A histogram can be used as a visual description of the frequency distribution of a continuous variable. We usually create between 5 and 20 class intervals where the endpoints of each class are not among the scores of the ungrouped frequency distribution. The range is the difference between the highest and lowest values in the distribution and is used to determine the width of the class intervals.

In Chapter 3 we shall explore ways in which we can use certain numerical techniques to examine the center of the distribution and the spread of the scores.

2.7 Problem Set

2.7.1 Estimates for 1972 (provided by the National Safety Council and the F.B.I.) on dangers we face are as follows:

Danger	Number of deaths
Death at the hand of strangers	5,000
Death from choking	3,900
Death by falling	17,000
Death by auto accidents	56,000

Source: Maggie Scarf, "The Anatomy of Fear," *New York Times Magazine,* June 16, 1974.

Create a bar graph of the frequency and relative frequency distribution of these data. Can you guess what Scarf's article is about?

2.7.2 Without looking at a clock, guess when 30 seconds have passed.
  (a) Create a frequency distribution of the times, to the nearest second, guessed by the class.
  (b) What per cent of the group was within 1 second of 30 seconds; that is, from 29 to 31 seconds?
  (c) What proportion of the group guessed 27 seconds or less?

(d) What per cent of the group guessed between 27 and 29 seconds (inclusive)?

(e) Create a bar graph of the distribution.

2.7.3 A record store sold 30 albums one evening. The following data represent the prices paid for each album: 1.89, 2.89, 2.89, 4.89, 3.89, 2.29, 3.29. 5.89, 1.89, 4.29, 2.89, 5.29, 4.29, 5.89, 3.89, 2.89, 3.29, 5.29, 6.29, 5.29, 4.29, 4.29, 1.89, 4.89, 3.89, 3.29, 5.29, 2.29, 2.89, and 6.29.

(a) Create a frequency distribution and bar graph to depict the sales distribution.

(b) Create a relative frequency and cumulative relative frequency column.

(c) What proportion of sales was at most $5.29?

(d) What per cent of sales was at least $4.89?

(e) What proportion of sales was between $3.89 and $5.89 (inclusive)?

2.7.4 If everyone in your class picked a number from 1 to 10, would you think that each number would be chosen approximately the same number of times? Try it. Create a histogram using class intervals of 0.5 to 1.5, and so on.

2.7.5 Create a histogram from the data in problem 2.7.2 using a class width of 2 seconds. (The boundaries of the class intervals should not be the same as any of the raw data.)

2.7.6 Draw ogive curves for the distributions in

(a) Problem 2.7.4.

(b) Problem 2.7.5.

2.7.7 How would you classify the distributions of

(a) Figure 2.2.1?

(b) Problem 3.6.1?

(c) Table 3.9.4?

2.7.8 Use a class width of 2 to graph the rest pulse data and compare them with your findings in Problem 2.3.4. What advantages or disadvantages do you see in comparison? Repeat the procedure and compare it with Problem 2.3.5.

2.7.9 After you look at the following data on cigarette brands, decide if they represent a sample or the population. If they represent a sample, should it be a random sample? Are the data comparable? What change would have to be made in the advertisement so that the data would be honestly presented?

Here's How the U.S. Government Ranks All These Cigarette Brands. (Compare Your Brand with Kent Golden Lights.)

Filter brands	Type	Tar (mg/cig)	Nicotine (mg/cig)
Kent Golden Lights	king size	8 mg	0.7 mg*
Merit	king size	9 mg	0.7 mg*
Vantage	king size	11 mg	0.7 mg
Multifilter	king size	13 mg	0.8 mg
Winston Lights	king size	13 mg	0.9 mg
Marlboro Lights	king size	13 mg	0.8 mg
Raleigh Extra Mild	king size	14 mg	0.9 mg
Viceroy Extra Mild	king size	14 mg	0.9 mg
Parliament	king size box	14 mg	0.8 mg
Doral	king size	15 mg	1.0 mg
Parliament	king size	16 mg	0.9 mg
Viceroy	king size	16 mg	1.1 mg
Raleigh	king size	16 mg	1.1 mg
Virginia Slims	100 mm	16 mg	1.0 mg
Parliament	100 mm	17 mg	1.0 mg
L&M	king size box	17 mg	1.1 mg
Silva Thins	100 mm	17 mg	1.3 mg
Marlboro	king size box	17 mg	1.0 mg
Raleigh	100 mm	17 mg	1.2 mg
Marlboro	100 mm	17 mg	1.1 mg
Benson & Hedges 100's	100 mm	18 mg	1.1 mg
Viceroy	100 mm	18 mg	1.2 mg
Marlboro	king size	18 mg	1.1 mg
Lark	king size	18 mg	1.2 mg
Camel Filters	king size	18 mg	1.2 mg
Eve	100 mm	18 mg	1.2 mg
Winston	100 mm	18 mg	1.2 mg
Winston	king size box	18 mg	1.2 mg
Chesterfield	king size	19 mg	1.2 mg
Lark	100 mm	19 mg	1.2 mg
L&M	king size	19 mg	1.2 mg
Tareyton	100 mm	19 mg	1.4 mg
Winston	king size	19 mg	1.3 mg
L&M	100 mm	19 mg	1.3 mg
Pall Mall	100 mm	19 mg	1.4 mg
Tareyton	king size	21 mg	1.4 mg

Source: FTC Report Apr. 1976.
* By FTC Method.

2.7.10 Given the following data on the mileage of small economy cars, where would the first car listed in the table appear if one drove 90 per cent of the time in the city and 10 per cent of the time on a highway? If the driving percentages were reversed, where would the first car appear in the list?

Make and model	Engine	Transmission	EPA estimates Highway City	
Pontiac Astre 2-door coupe	140-cu.-in. 4-cyl. 2-bbl.	Manual	35	22
Toyota Corona 2-door sedan	133.6-cu.-in. 4-cyl. 2-bbl.	Manual	34	20
Datsun 710 2-door sedan	119-cu.-in. 4-cyl. 2-bbl.	Manual	33	23
Fiat 131 2-door sedan	107-cu.-in. 4-cyl. 2-bbl.	Manual	29	18
Mazda RX-3 coupe	70-cu.-in. rotary 4-bbl.	Manual	30	19
VW Dasher 2-door sedan	97-cu.-in. 4-cyl. F.I.	Manual	37	24
Volvo 242 2-door sedan	130-cu.-in. 4-cyl. F.I.	Manual	27	17

Source: 1976 EPA Fuel Economy Guide.

Note: Remember that these mileage figures are estimates. The mileage you get will vary according to the kind of driving habits, and your car's condition and available equipment.

2.7.11 If heating costs for a community during the month of February varied from $27.54 to $116.28, construct a table with six classes into which these amounts may be grouped.

2.7.12 Given the following class marks—28, 43, 58, 73, and 88—create the class intervals.

2.7.13 From the following graphs determine:
  (a) The normal range of height for boys aged: 6 to 8, 12 to 14.
  (b) The normal range of height for girls aged: 6 to 8, 12 to 14.
  (c) The normal range of weight for boys aged: 8 to 10, 14 to 16.
  (d) The normal range of weight for girls aged: 8 to 10, 14 to 16.
  (e) The height of boys aged 12 who fall to the right of the normal range; that is, boys who are taller than the maximum height of the normal range.
  (f) The answer to part (e) for girls aged 12.
  (g) The lowest weight of boys aged 16 that falls above the normal range.

(h) The highest weight of girls aged 16 that falls below the normal range.

(i) If a girl aged 8 whose height is 44 inches would be referred to a doctor.

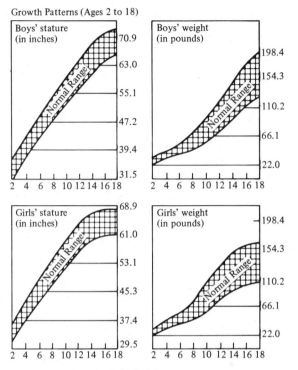

Growth Patterns (Ages 2 to 18)

Source: National Center for Health Statistics

NOTE: The charts show the height and weight within which 80 per cent of growing children fall. Children above or below these limits might be referred to doctors to see if there is a medical cause.

2.7.14 The following are the number of napkins a restaurant used on 50 consecutive Sundays:

```
37  48  42  38  42  32  40  64  54  40  45  44  44  61
48  46  41  68  41  39  43  45  53  40  44  45  41  43
44  39  41  52  57  50  47  43  38  49  37  41  54  51
39  59  34  38  39  42  45  43
```

(a) Group these figures into a table having the classes 31 to 35, 36 to 40, 41 to 45, etc.

(b) Convert the distribution in (a) into a cumulative frequency distribution.

(c) If the manager of the restaurant decided to stock only 55 napkins on each Sunday, what per cent of the time could she expect to run out of napkins?

(d) Draw a histogram of the distribution found in (a).

(e) Draw an ogive of the cumulative percentage distribution, and determine approximately below which value 75 per cent of the data falls.

Chapter 3

Measures of Central Tendency and Variability

3.1 Introduction

In Chapter 2 we discussed how creating tables and graphs helps to organize data. Graphic representations, in terms of a frequency table or a histogram, give us a greater understanding of the data distribution. What is even more helpful is to find a few numerical values that express the distribution of the data. In this chapter we shall develop two such concepts. One is a measure of the central tendency of a set of data (average) and the other is a measure of the variability of the data (spread of the data around the average). Statistically speaking, although a picture (here, a frequency table or graph) may be worth a thousand words, a few numbers almost take its place.

Modern society is very number-oriented. In newspaper and television reporting, averages are often used as a shorthand form of representing trends, such as baseball batting averages, the average cost-of-living increase, or the average children per family. An *average* is the number that represents the central tendency of the data. In statistics, three measures of average or central tendency are frequently used: the mode, the median, and the arithmetic mean. We shall look at all three averages and examine situations in which one of them is more useful than the others in describing a quantitative situation. In addition, inasmuch as the average of a set of data does not tell us how the data are distributed, we shall also examine measures of variability about the averages.

3.2 The Mode

The *mode* is the number that occurs most frequently in a set of data. Suppose that the manager of a bus company wanted to determine if a certain stop on the bus route should be continued. The number of people getting off at the stop in 11 trips produced the following data: 2, 3, 7, 9, 9, 10, 12, 12, 12, 14, and 15—the mode is 12. Inasmuch as in 11 different trips the number of passengers getting off at this stop with greatest frequency was 12, it appears that this stop is frequently used and therefore should be included in the bus route.

The mode is easy to determine once the data have been organized. To organize the data, the scores should be put either in ascending (from the smallest to largest) or descending order (from largest to smallest). In this example, the data were presented in ascending order. This measure is used when it is important to know which value occurs most frequently.

As a descriptive statistic, the mode is of limited value. If there are a large number of data, their organization is a time-consuming process. More important, however, the mode does not tell us anything else about the distribution of the rest of the scores. For example, the mode of the bus-stop distribution would still be 12 even if the data were distributed in the following manner: 0, 0, 1, 1, 2, 3, 4, 5, 12, 12, and 12. For this reason, the mode has little value in describing the distribution of scores. If the data had been 0, 0, 1, 2, 3, 4, 7, 8, 10, 12, and 12, the distribution would be referred to as *bimodal*; that is, there are two (bi) values that occur most frequently. You can see that a decision to keep the bus stop based only on the value of the mode is questionable.

3.3 The Median

Another measure that is referred to as an average is the *median*. The median is a number in a set of ordered data above which 50 per cent of the scores fall and below which 50 per cent of the scores fall.

METHODS FOR DETERMINING THE MEDIAN

There are essentially three methods in use for determining the median. The best method to use depends upon whether the ordered data consist of (a) an odd number of scores, (b) an even number of scores, or (c) class intervals.

Odd Number of Data

When there are an odd number of ordered data, the median is the middle number. Consider the scores 2, 3, 7, 9, 9, 10, 10, 12, 12, 12, 14, and 15; the median is 10. Since there are 11 scores, there are 5 scores above 10 (12, 12, 12, 14, 15) and 5 scores below 10 (2, 3, 7, 9, 9). In other words, half of the data are above 10 and half of the date are below 10.

Even Number of Data

If we join 113 to the scores, we have 2, 3, 7, 9, 9, 10, 12, 12, 12, 14, 15, and 113—the median now lies between the 10 and the 12. Any value between 10 and 12 would divide the data so that 50 per cent of the scores are below it and 50 per cent of the scores are above it. It is common practice to take the midpoint of the two "middle" scores when finding the median of an even number of data. In this case we would find the value between 10 and 12 by adding them and dividing by 2. Hence $\dfrac{10+12}{2} = 11$ is the median score for 2, 3, 7, 9, 10, 12, 12, 12, 14, 15, and 113.

Class Intervals

When data have been put into a frequency distribution with class intervals (Table 3.3.1), the median is that score which divides the histogram of the data (Figure 3.3.1) into two equal areas.

To determine the median in a histogram, we find the value that lies above and below 50 per cent of the scores. In Figure 3.3.1, the histogram shows a total frequency of 60 scores. We locate a value that is above 30 of

Table 3.3.1. Frequency and Class-Interval Distribution

Score	Frequency	Score	Frequency	Class interval	Frequency
48	1	70	5	47.5–52.5	3
50	1	71	2	52.5–57.5	5
51	1	72	1	57.5–62.5	8
53	1	73	3	62.5–67.5	12
54	1	74	1	67.5–72.5	10
56	2	75	2	72.5–77.5	9
57	1	76	3	77.5–82.5	6
58	1	78	1	82.5–87.5	5
60	4	79	1	87.5–92.5	1
61	1	80	3	92.5–97.5	1
62	2	81	1		60
63	2	83	1		
64	3	84	1		
65	3	85	2		
66	2	86	1		
67	2	90	1		
68	2	96	1		
			60		

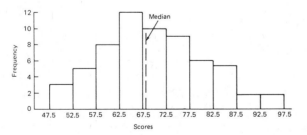

Figure 3.3.1. *Histogram*

the scores and below 30 of the scores. There is a frequency of 28 (3 + 5 + 8 + 12) in the first four class intervals. Hence the median lies somewhere in the fifth class interval: 67.5 to 72.5. We need a frequency of 2 to add to the 28 to create a frequency of 30. Since 2 of the 10 scores in the interval 67.5 to 72.5 are required, we locate the point that is $\frac{2}{10}$ of the interval of class width 5. The median is therefore $67.5 + \frac{2}{10}(5) = 67.5 + 1 = 68.5$.

This value, 68.5, differs from the median found for the raw data, which was 69 (Table 3.3.1). The median for raw data and the median for a histogram usually differ. This is due to the fact that we have assumed that in each class interval the scores are distributed equally, and this is generally not the case in histograms. For example, in the interval 67.5 to 72.5, there are 10 scores, but the 10 scores are not distributed equally in the class interval: there are two 68's, five 70's, two 71's, and one 72. However, the approximation is usually quite close.

To find the median for data grouped into class intervals, determine:

1. The lower bound, l, of the class interval containing the median (67.5 for our example).
2. The width, w, of the interval (5).
3. The frequency, f, needed to reach a frequency of $\frac{n}{2}$ ($n = 60$; $f = \frac{n}{2} - 28 = 30 - 28 = 2$).
4. The number, n, of scores in the interval (10). Then the median value is: $m = l + \frac{f}{n}(w)$, or $67.5 + \frac{2}{10}(5) = 68.5$.

The median is used when a "typical" score is required (a score that falls in the "middle" of the ordered data). It is not altered significantly by the presence of an extreme value. For example, compare the medians found for an odd and an even number of data above when 113 was included in the data. Like the mode, the median has limitations. When a number of samples are taken from a population, the medians from the samples can vary considerably. For example, suppose that the population scores are the digits: 0, 1, 2, ..., 9. A few samples of size $n = 3$ would be 2, 4, 6; 1, 2, 9; 3, 5, 8; and 2, 9, 9. Each of these samples has a different median value. Therefore, the median lacks the necessary stability to be used in inferential statistics and is used primarily in descriptive statistics. Recall that in the bus-stop problem the original data has a median of 10 and a mode of 12. In the second set of data the median is 2.5 but the mode is still 12. You can see that the median provides a different picture of the data than the mode.

P R O B L E M S

3.3.1 (a) Find the mode and median of the following 29 pieces of data: 23, 10, 17, 20, 18, 13, 20, 15, 16, 11, 9, 14, 10, 12, 7, 9, 23, 14, 19, 22, 11, 15, 7, 23, 11, 15, 9, and 17.

(b) Organize the data into class intervals and find the modal class (the class containing the greatest frequency) and the class-interval median.

3.3.2 Find the modal class and the class-interval median of the data given in Figure 2.3.1.

3.3.3 Find the median and mode of your:
(a) Rest pulse data.
(b) Jogging pulse data.

3.3.4 What are the model classes and medians of the distributions in Problem 3.3.3, where the class widths of 2 and 4 were used?

3.4 The Mean

The average most of us are familiar with is the *arithmetic mean*. It is the measure of central tendency most often used to determine average expense, average grade, average time, and so on.

METHODS FOR OBTAINING THE MEAN

There are three methods for determining the mean, depending upon whether the data are (a) unorganized, (b) in a frequency distribution, or (c) in class intervals.

Unorganized Data

To determine the mean of a set of raw data, we add up the scores and divide by the total number of scores. For example, if the scores are 10, 9, 7, 3, 12, 14, 12, 15, 2, 12, and 9, the mean is

$$\frac{10+9+7+3+12+14+12+15+2+12+9}{11} = \frac{105}{11} = 9.\bar{5}$$

If the data represent the entire *population* data, the mean is written as μ (read: mew). To show that this mean is for a set of scores labeled x-scores, use μ_x. If the data are of a *sample* taken from a population of x-scores, use $\bar{x}$ (read x bar) for the mean. Thus, if the data are from a sample, the mean,
$\bar{x} = \frac{\text{sum of scores}}{\text{total number of scores}}$. If the data represent the entire population, replace $\bar{x}$ with μ_x. We shall use n when we are referring to sample size, and N when we are referring to population size.

In general, if the first piece of data is denoted by x_1, the second by x_2, and so on, then the mean of n scores is either

$$\bar{x}(\text{sample mean}) = \frac{x_1 + x_2 + \cdots + x_n}{n}$$

or

$$\mu_x \text{(population mean)} = \frac{x_1 + x_2 + \cdots + x_N}{N}$$

Instead of writing "$x_1 + x_2 + x_3 + \cdots + x_{11}$," we can use mathematical shorthand. The notation $\sum_{j=1}^{5} j$ (read: sigma j from $j = 1$ to 5) is used to denote the sum $1 + 2 + 3 + 4 + 5$. Similarly, $\sum_{i=1}^{11} x_i$ is equivalent to writing "$x_1 + x_2 + \cdots + x_{11}$." (For a more detailed account of the summation symbol, see Appendix A.) Thus, using this notation, the mean can be written

$$\bar{x} = \frac{\sum_{i=1}^{n} x_i}{n} \quad \text{or} \quad \mu_x = \frac{\sum_{i=1}^{N} x_i}{N}$$

Frequency Distribution

When the data have been put into a frequency table (Table 3.3.1), the mean can be found by grouping the scores and adding them up using their frequencies of occurrence; that is, instead of writing 73 three times in the sum, we can write it as 3×73 (Table 3.3.1). Then

$$\bar{x} = \frac{48 + 50 + \cdots + 2 \cdot 56 + \cdots + 4 \cdot 60 + \cdots + 3 \cdot 73 + \cdots + 96}{60} = \frac{4{,}145}{60}$$

$$= 69.08$$

The result is the same as that obtained by adding all the scores and dividing by 60. Instead of writing all 60 pieces of data, we find the sum using the 34 different values and their corresponding frequencies.

In general, if the data are grouped into k different values or classes, so that f_1 is the frequency of the x_1 score, f_2 is the frequency of the x_2 score, ..., and f_k is the frequency of the f_k score, then the sample mean

$$\bar{x} = \frac{x_1 f_1 + x_2 f_2 + \cdots + x_k f_k}{n} = \frac{\sum_{i=1}^{k} x_i \cdot f_i}{n}$$

(Note that the denominator is still n, the number of pieces of data.)

Class Intervals

When the data have been put into class intervals, the mean is determined by using the class marks of the intervals and the corresponding frequencies of the class intervals (Table 3.4.1). That is, the class marks are multiplied by their corresponding frequencies and summed to create a total score. The total score is then divided by n, the number of scores.

Table 3.4.1

Class intervals	Class mark	Frequency
47.5–52.5	50	3
52.5–57.5	55	5
57.5–62.5	60	8
62.5–67.5	65	12
67.5–72.5	70	10
72.5–77.5	75	9
77.5–82.5	80	6
82.5–87.5	85	5
87.5–92.5	90	1
92.5–97.5	95	1

$$\bar{x} = \frac{(3 \times 50) + (5 \times 55) + (8 \times 60) + \cdots + (1 \times 90) + (1 \times 95)}{3 + 5 + 8 + \cdots + 1 + 1} = \frac{4,150}{60} = 69.16$$

The formula is the same as that used when the data have been grouped, except that here k represents the number of class intervals.

The value of $\bar{x}$ for class intervals is close to the value of $\bar{x}$ when raw data were used. The difference is due to the assumption of an equal distribution in the class intervals. We observed a similar situation in the case of the median.

The mean has the advantage of not requiring the data to be organized, whereas the mode and the median can only be obtained from ordered data. We saw that when we joined an extreme value to the data, the mode and the median were not affected significantly, if at all. This is not the case with the mean. If we now use $x_{12} = 113$ in our data, then the mean

$$\bar{x} = \frac{10 + 9 + 7 + \cdots + 9 + 113}{12} = \frac{218}{12} = 18.17$$

Thus, the mean can be altered considerably by extreme values. However, when taking samples from a population, the mean does not vary as much as the median. For example, earlier we took 4 samples of $n = 3$ from the digits. The corresponding medians are 4, 2, 5, and 8. The corresponding means are 4, 4, 5.33, and 6.17. Mean sample scores vary less in general than median sample scores, and thus the mean is used widely in inferential statistics.

3.5 Choosing an Average

Sometimes one measure of central tendency is more informative than the others, and sometimes one can be more misleading than the others. For example, to attract people a company may advertise that the average employee's salary is $16,250. If many of the employees earned around $10,000 but one of them earned $50,000, then, using the arithmetic mean, the average salary could be $16,250. If the median had been used, the average salary would be $12,500, whereas if the mode had been used, it would be $10,000. Situations similar to this probably helped foster the idea that there are three types of lies: "lies, damned lies, and statistics."

The mode is the only average that is guaranteed to be one of the observed values. The median and the mean can fall between two observed values and not be one of the pieces of data. The mode does not tell us anything about the distribution other than the value that occurred most often. The median is a good measure to use when a "typical" individual or score is required. The value of the median lies in the fact that it gives insight into the distribution by determining that 50 per cent of the scores were less than it and 50 per cent were greater. Do you know what the divider between the opposing lanes on highways is called? Can you see why this choice was made? In the bus-stop problem, the median is seen to be more valuable than either the mode or mean, since the mean can vary greatly when an extreme score is introduced, and the mode does not give us enough information about the entire distribution—which is important in this situation. Can you think of a situation in which the number of people getting off at a usually unused bus stop could be large?

Since one of our purposes is to examine samples and reach conclusions about the populations from which they came (i.e., inferential statistics), the mean is the measure of central tendency on which we shall rely heavily. This choice results from the fact that the mean has greater stability than either the mode or median. That is, when we take samples from a population, the sample means vary less than the sample modes or medians.

P R O B L E M S

3.5.1 The data presented here represent the number of hours spent studying for an examination by 20 students in a mathematics class: 4, 5, 3, 6, 7, 1, 2, 3, 0, 5, 6, 5, 8, 4, 0, 2, 3, 7, 5, and 6.
   (a) Find the mode.
   (b) Find the median.
   (c) Find the mean.
   (d) Classify the data into a frequency table.
   (e) Draw a histogram (use nine intervals).
   (f) Calculate the modal class.
   (g) Calculate the class-interval median.
   (h) Calculate the class-mark mean.

3.5.2 *Class Project*: Determine the mean of the first 250 two-digit numbers in the random numbers table, in Table VII in Appendix B. Then choose 5 *random* samples of 10 two-digit numbers and determine the 5 sample means. Do this again with 5 *random* samples of 20 two-digit numbers. What conclusions can the class draw?

3.5.3 If the mean income for 20 analysts is $17,000, what is the total income of the group?

3.5.4 The mean test score for the 20 students in class 204 was 74. The mean for 18 students in another section was 81. What was the mean for the combined group?

3.5.5 *Comment*: A teacher gave a standardized test to each of his five classes. From the data he determined the five median scores. He averaged them and used this value as an average measure of his classes' ability.

3.5.6 At a meeting of dumpling makers 20 persons were asked to indicate their annual salary. The distribution is as follows:

Annual salary	Frequency
$5,500	7
6,000	5
7,000	6
8,000	4
30,000	3
	20

Which measure of central tendency would you use to report the data? Why would you not choose either of the other two averages?

3.5.7 Create four distributions each containing 20 scores, where the mean of each is 8.

3.5.8 *Class Project*: Combine the distributions created by everyone in the class in response to Problem 3.5.7 into a single frequency distribution where the x-score represents a particular distribution; for example, how many people chose a distribution where there were 19 scores of 0 and 1 score of 160, which yields a mean = 20. What kinds of inferences can you make about the students in your class when you look at the particular distributions, and their frequencies, that they chose? For example, does it appear that your group has an affinity toward symmetry? Even numbers?

3.5.9 *Class Project*: Using a graduated cylinder, determine the liquid measure of 10 bottles of soda that claim 12 ounces of liquid; then determine the three averages. What does each of the averages tell you?

3.5.10 If someone invests $6,000 at 5 per cent interest, $7,500 at 7 per cent, and $8,000 at 8 per cent, what is the mean interest for these investments?

3.5.11 Refer to Problem 14 at the end of Chapter 2 and determine:
   (a) The mean of the ungrouped data.
   (b) The mean of the grouped data.
   (c) The number of napkins the manager should have available so that she can expect to run short only 10 per cent of the time.
   (d) The number of napkins the manager should have so as to have 36 or more napkins left over only 25 per cent of the time.

3.6 Introduction to Variability

When we hear of someone who is ill with a high or low temperature, or someone who did exceptionally well or poorly on an examination, or someone who makes a very high salary or a very low salary, this information catches our attention. In these situations the particular numerical value differs considerably from what we would expect; that is, there is considerable variation between what is the norm and the particular information. Earlier we had mentioned that certain groups of people such as athletes and astronauts have considerably lower pulse rates than the "average" person; similarly, people who practice transcendental meditation or bio-feedback are able to control their brain activity and spend considerably more time each day in a relaxed state than most of us. These groups of people are of interest to us because of the variation of their condition compared to the population at large. In general, it is important in examining data to find the average, but it is just as important to know about the variation about the average.

The question that economists, medical people, sports fans, teachers, and in fact, any group with special interests must consider involves determining what situations will be labeled extreme and also whether the extreme situation or data can be explained as due to chance alone or due to some quality or condition. For example, if someone has a wound that is healing, doctors watch the wound to determine if the wound is healing approximately as the average person would heal. Some people heal faster or slower than others. In this situation doctors would have to be concerned about slow healing. The question is, how slow is slow? That is, they must decide if the person just heals somewhat slower than the average person or if, in fact, the person has an infection that is preventing the healing. In sports the situation is quite similar. For example, suppose that a player is a star player based on past observation. Then one day she has a bad day. The question is, is this variation just due to chance; that is, has she just had one of those bad days we all have or does it reflect something else, namely, is she getting older or is she hurt? How many times would we have to observe poor performances before we would begin to think that the variation from the normal performance is significant? Can you think of other situations in which variations from what is expected has been your concern? The concept of variability about the average is an important consideration, and we shall consider different ways of looking at variability in the next two sections.

PROBLEMS

3.6.1 Find three situations in your life in which you were concerned with variations from what was expected.

3.6.2 Find three situations in your life in which you were not concerned with variation from what was expected.

3.6.3 If your lunch usually costs on the average of $1, how would you feel if it were suddenly to cost 95 cents, $1.05, 88 cents, $1.17, 58 cents, or $1.77?

3.7 Some Measures of Variability

The measure of variation that is simplest to determine is the range. The range of a set of scores is the difference between the largest and smallest scores in the set of scores. For example, if the smallest score of a collection of scores is 10, and the largest score is 73, the range is $73 - 10 = 63$. You can see that although this statistic is easy to determine, it is of very limited value. For example, the range of 63 would still be the range no matter how many scores were collected as long as 10 and 73 were the smallest and largest scores. That is, the range tells us about the extreme scores but does not tell us about the variation among all the scores. For this reason it is used in statistical situations to give a "feeling" of the scores rather than an understanding.

PROBLEMS

3.7.1 If you were going to eat in a restaurant, which of the following price ranges would you choose (and why?): $2 to $10, $5 to $10, or $8 to $10.

3.7.2 Suppose that three companies had a range of sales, in one month, of $30,000. Is this information complete enough to make a general statement about the three companies? Why?

3.7.3 Suppose that the same three companies had as their lowest sale $2,000, and $32,000 as their highest sale. Sketch three sales distributions, having the same range, but whose distributions are considerably different. Assume that each company made 20 sales in the month.

In Section 3.3, we examined the median. The median is often referred to as the 50th percentile ($P_{.50}$), since half the scores are below it and half the scores are above it. To discuss the variability about the median, we can examine the 25th ($P_{.25}$) and 75th ($P_{.75}$) percentiles. These particular percentiles are also known as quartiles: In a histogram, 25 per cent of the scores lie below the 25th percentile, and 25% of the scores lie above the 75th percentile. Thus if we determine the 25th percentile (Q_1, the first quartile), the 50th percentile (the median—also referred to as the second quartile), and the 75th percentile (Q_3, the third quartile), we shall obtain the average and the variability about the average. The variation about the median is known as the *interquartile range* and is the range of the middle 50 per cent of the scores; the interquartile range $= Q_3 - Q_1$.

The formulas for the 25th and 75th percentiles are similar to the formula for the median; namely, $m = l + f(w)/N$. In an earlier problem (Table 3.4.1), the number of pieces of data, n, was 60. The 25th percentile would be in the class interval containing the 15th score (25 per cent of 60). The formula is: $Q_1 = l + f(w)/N$, where l is the lower bound of the interval containing the 15th score; N is the number of scores in the interval; w is the width of interval; and f the frequency needed to reach the $N/4 =$ the 15th score. We find the 25th percentile in the 3rd interval 57.5 to 62.5 (Figure 3.3.1). Since there is a frequency of 8 in the preceding intervals, $f = 15 - 8 = 7$, $w = 5$, and $N = 8$. So, $Q_1 = 57.7 + \frac{7}{8}(5) = 57.5 + 4.4 = 61.9$. You should make sure that you can determine Q_3; we found it to be 76.4. Then we can determine the interquartile range to be $76.4 - 61.9 = 14.5$. That is, the middle 50 per cent of the scores fall within a range of 14.5. In addition to the first and third quartiles, we can find any percentile we wish to determine. For example, we can find the 95th percentile, the score above which 5 per cent of the scores lie; namely, $P_{.95} = 82.5 + \frac{4}{5}(5) = 86.5$. Check us to see if this score is correct.

P R O B L E M S

3.7.4 Find Q_1, Q_2, and Q_3, and the interquartile range of the data in Table 3.6.2.

3.7.5 Find $P_{.05}$ and $P_{.95}$ of the data in Table 3.6.2.

3.7.6 Can you create a general formula to determine any percentile?

As we have just seen, the interquartile range describes only a particular variation measure and does not tell us about the entire distribution. Additionally, the use of this measure is, or should be, restricted to situations in which there are a considerable number of scores, around at least 100. If there are only a few scores, then the problem appears at the point where a particular score can be said to represent a wide range of percentiles. Another problem occurs when two different rankings could in fact be used for two scores that are quite close numerically. Because of these problems percentiles are usually used in situations where there is a considerable number of scores such as found in achievement test scores collected nationally, and for these reasons percentiles are not used frequently in the study of statistical inference. The measure of variation most frequently used in the study of statistical situations is the standard deviation—the measure of variation about the mean.

3.8 Standard Deviation

It is necessary to obtain a measure of variation about an average that can be used to tell us about the distribution of scores about the average. In determining the mean of a set of scores, we locate where the data are centrally located, but it does not tell us how the rest of the scores disperse or spread about the mean. For example, the average grade on an examination in a psychology class may be the same for all the psychology classes taking that test, but we do not know, and cannot assume, that the total distribution of scores was also the same (Table 3.8.1).

Table 3.8.1. Frequency Distribution of Grades

Score	Frequency of one class	Frequency of all classes
50	0	10
60	4	10
70	8	25
80	4	45
90	4	5
100	0	5
	$\overline{20}$	$\overline{100}$

$\bar{x}$ (sample of one psychology class)

$$= \frac{60 \cdot 4 + 70 \cdot 8 + 80 \cdot 4 + 90 \cdot 4}{20} = \frac{1,480}{20} = 74$$

μ_x (population of all psychology classes)

$$= \frac{50 \cdot 10 + 60 \cdot 10 + 70 \cdot 25 + \cdots + 100 \cdot 5}{100} = \frac{7,400}{100} = 74$$

Notice that the scores for the one class went from 60 to 90, whereas for all the classes the distribution included scores as low as 50 and as high as 100. We observe that one class had a range of $90 - 60 = 30$ points, whereas the total group had a range of $100 - 50 = 50$ points. The range does vary with sample size; it can only get bigger the more values that are chosen. (Why?) Additionally, it is very sensitive to extreme scores. (Why?)

We could also have two distributions that agree on the value of the mean and have the same number of measurements but have different distributions. For example, if the x-scores were 2, 4, 6, 8, and 10, the mean would be 6; and if the y-scores were 1, 8, 12, 7, and 2, the mean would also be 6. Although both have the same number of scores, 5, the distributions are clearly different.

If we graphed the two distributions using a solid dot to represent x-scores and an open circle to represent y-scores, we notice the differences in variability (Figure 3.8.1). The y-scores have a greater dispersion than the

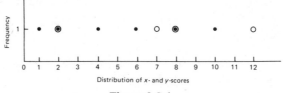

Distribution of x- and y-scores

Figure 3.8.1

x-scores. We shall assume that the x-scores are the total population scores; the mean is $\mu_x = 6$. We want to measure the dispersion from μ_x. It is natural to try as a measure of dispersion the sum of the differences between the individual scores and the mean:

$$(x_1 - \mu_x) + (x_2 - \mu_x) + (x_3 - \mu_x) + (x_4 + \mu_x) + (x_5 - \mu_x)$$

Thus we have

$$(2-6) + (4-6) + (6-6) + (8-6) + (10-6) = -4 + -2 + 0 + 2 + 4 = 0$$

The sum of the differences from the mean will always be zero. Why? This characteristic of the mean suggests that the mean is the balancing point of the distribution.

P R O B L E M

3.8.1 (a) What is the range of the x-scores?
(b) What is the range of the y-scores?
(c) Show that the sum of the differences from the mean for the y-scores is also zero.

We now take another approach to the problem of getting a measure of variability. We first square the difference between each score and the mean (this eliminates negative values) and then sum these squares of differences (Table 3.8.2). Thus the sum of the squared deviations from the mean:

$$\sum_{i=1}^{5} (x_i - \mu_x)^2 = 40.$$

Table 3.8.2

x	$x - \mu_x$	$(x - \mu_x)^2$
2	-4	16
4	-2	4
6	0	0
8	2	4
10	4	16
	$\overline{0}$	$\overline{40}$

To obtain the average of the squared deviations from the mean, we divide by $N = 5$. For a population distribution, the average of the squared deviations from the means is known as the *variance* and is written as σ^2 (read: sigma squared):

$$\sigma^2 \text{ (variance)} = \sum_{i=1}^{5} \frac{(x_i - \mu_x)^2}{5} = \frac{40}{5} = 8$$

If the x-scores were in units of pounds, inches, or any other unit, squaring the differences creates units of square pounds, square inches, or whatever. To change back to the original units, we take the square root of the variance. The square root of the variance $\sigma^2 = \sqrt{\sigma^2} = \sigma$ and is called the *standard deviation*. That is,

$$\sigma = \text{(standard deviation)} = \sqrt{\frac{\sum_{i=1}^{N} (x_i - \mu_x)^2}{N}} = \sqrt{8} = 2.83$$

Since the mean, μ_x, equals 6, in the interval of 1 standard deviation (2.83 units) on each side of the mean, there are three scores: 4, 6, and 8. That is, 1 standard deviation to the left of the mean is $\mu_x - 1\sigma = 6 - 2.83 = 3.17$;

and 1 standard deviation to the right of the means is $\mu_x + 1\sigma = 6 + 2.83 = 8.83$. The interval of $\pm 1\sigma$ (plus or minus 1 standard deviation) from the mean creates an interval of $(\mu_x - 1\sigma, \mu_x + 1\sigma) = (3.17, 8.83)$. In this interval we find the scores 4, 6, and 8. Since there are five scores in the distribution, $\frac{3}{5} = 60$ per cent of the scores lie within 1 standard deviation of the mean.

P R O B L E M

3.8.2 (a) Show that 100 per cent of the scores are within 2σ's from the mean. That is, show that the interval $(\mu_x - 2\sigma, \mu_x + 2\sigma)$ contains all the x-scores.

(b) What score lies exactly 1.5 standard deviations above the mean? Below the mean?

There are two formulas for determining the standard deviation of population data, depending upon whether the data are ungrouped or grouped.

1. Ungrouped data:

$$\sigma_x = \sqrt{\frac{\sum_{i=1}^{N} (x_i - \mu_x)^2}{N}} \qquad \mu_x = \text{mean}, \quad N \text{ pieces of data}$$

2. Grouped data: Where there are k groups with N scores and frequencies $f_1, f_2, \ldots, f_k$.

$$\sigma_x = \sqrt{\frac{\sum_{i=1}^{k} (x_i - \mu_x)^2 f_i}{N}}$$

We saw that the variability of y-scores about the mean was greater than that for the x-scores. Suppose that the y-scores are a sample taken from a population; the mean would be $\bar{y} = 6$. We now find the sample standard deviation for the y-scores. The y_i, $y_i - \bar{y}$, and $(y_i - \bar{y})^2$ values are shown in Table 3.8.3. The sum of the squared deviations $\sum_{i=1}^{5} (y_i - \bar{y})^2 = 82$. This is greater than for the x-scores, as we would expect.

To find the average squared deviation for a sample, the *sample variance*, s^2, we do not divide by N but by $n - 1$. That is,

$$s^2 \text{ (sample variance)} = \frac{\sum_{i=1}^{5} (y - \bar{y})^2}{n - 1} = \frac{82}{4} = 20.5$$

Table 3.8.3

y_i	$y_i - \bar{y}$	$(y_i - \bar{y})^2$
1	−5	25
8	2	4
12	6	36
7	1	1
2	−4	16
	0	82

The standard deviation for a sample, s, is found by taking the square root of the sample variance:

$$s = \sqrt{\frac{82}{4}} = \sqrt{20.5} = 4.5$$

Statisticians have found that dividing by $n-1$ gives a closer value of the standard deviation for the entire population. The precise reason for this is beyond the scope of this text.*

Since $\bar{y} = 6$ and $s = 4.5$, the interval $(6 - 4.5, 6 + 4.5) = (1.5, 10.5)$ represents 1 standard deviation from the mean. Thus, there are three scores, 2, 7, and 8, which fall within 1 standard deviation of the mean. That is, $\frac{3}{5} = 60$ per cent of the scores are within 1 standard deviation of the mean. Note that the standard deviation for the x-scores was 2.83, versus 4.5 for the y-scores.

There are two formulas for determining the standard deviation of sample data, depending upon whether the data are ungrouped or grouped. These formulas are the same as those for the two population standard deviations, except that s replaces σ and $n-1$ replaces n:

1. Ungrouped data:

$$s = \sqrt{\frac{\sum\limits_{i=1}^{n}(y_i - \bar{y})^2}{n-1}}$$

2. Grouped data:

$$s = \sqrt{\frac{\sum\limits_{i=1}^{k}(y_i - \bar{y})^2 \cdot f_i}{n-1}}$$

---

* See, for example, Paul G. Hoel, *Introduction to Mathematical Statistics*, 4th ed. (New York: John Wiley & Sons, Inc., 1971).

PROBLEM

3.8.3 Suppose that the y-scores were population scores.

(a) Compute σ_y. Is $\sigma_y > \sigma_x$?

(b) Find the percentage of scores within 1 standard deviation of μ_y.

(c) Find the percentage of scores within 2 standard deviations of μ_y.

When data have been put into class intervals, the mean and standard deviation, like the median and quartile scores, are approximations of the mean and standard deviation of the original data. For example, suppose that a group (population) of 10 seventh graders took the following time, in seconds, to climb a rope and descend: 18, 20, 23, 24, 27, 29, 32, 35, 39, and 43. We shall organize the data into class intervals, class marks, and their corresponding frequencies (Table 3.8.4), and determine μ_x and σ_x from the grouped data.

Table 3.8.4

Class interval	Class mark	Frequency
17.5–22.5	20	2
22.5–27.5	25	3
27.5–32.5	30	2
32.5–37.5	35	1
37.5–42.5	40	1
42.5–47.5	45	1

$$\mu_x = \sum_{i=1}^{k} \frac{x_i \cdot f_i}{10} = \frac{20 \cdot 2 + 25 \cdot 3 + 30 \cdot 2 + 35 \cdot 1 + 40 \cdot 1 + 45 \cdot 1}{10}$$

$$= 29.5 \text{ seconds}$$

To determine the standard deviation, we have to find the average squared deviations from the mean and then take its square root. The standard deviation can be determined as shown in Table 3.8.5.

$$\sigma^2 \text{ (variance)} = \frac{\sum_{i=1}^{6} (x_i - \mu_x)^2 \cdot f_i}{10} = \frac{622.50}{10} = 62.25$$

$$\sigma \text{ (standard deviation)} = \sqrt{62.25} \approx 7.9 \quad \text{(Table I in Appendix B)}$$

Thus

$$\mu - 1\sigma = 29.5 - 7.9 = 21.6 \quad \text{and} \quad \mu + 1\sigma = 29.5 + 7.9 = 37.4$$

Table 3.8.5

Class mark	Frequency	$x_i - \mu_x$	$(x_i - \mu_x)^2$	$(x_i - \mu_x)^2 \cdot f_i$
20	2	−9.5	90.25	180.50
25	3	−4.5	20.25	60.75
30	2	0.5	0.25	0.50
35	1	5.5	30.25	30.25
40	1	10.5	110.25	110.25
45	1	15.5	240.25	240.25
	10			622.50

It follows, then, that 1 standard deviation about the means is represented by the interval (21.6, 37.4). As

$$\mu - 2\sigma = 29.5 - 15.8 = 13.7 \quad \text{and} \quad \mu + 2\sigma = 29.5 + 15.8 = 45.3$$

2 standard deviations about the mean is represented by the interval (13.7, 45.3). Figure 3.8.2 shows the graphic interpretation of standard deviation.

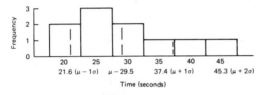

Figure 3.8.2

PROBLEMS

3.8.4 Find the mean and standard deviation for the ungrouped climbing-rope data (Table 3.8.4).

3.8.5 Find the mean and standard deviation for your rest pulse data.

3.8.6 Find the mean and standard deviation for your jogging pulse data.

3.8.7 Later we shall ask you to prove that the following statement is true:

$$\frac{1}{n} \sum_{i=1}^{n} (x_i - \bar{x})^2 = \frac{1}{n} \sum_{i=1}^{n} x_i^2 - \bar{x}^2$$

The right side of this equation provides an easier way to calculate the variance, and thus, the standard deviation. To convince yourself that it is easier, find the mean and the standard deviation for the following data using both variance formulas.

$$x_i: 2, 4, 6, 8, 10, 12$$

3.8.8 The formula given in Problem 3.8.7 is used when the data are ungrouped, or in another way to say it, where each score is its own interval. When the data are grouped, class intervals containing more than one score are considered so that it is necessary to introduce the frequencies of the intervals to determine the variance. As you have seen in the variance formula that followed Table 3.8.5, the

$$\text{variance} = \frac{1}{n} \sum_{i=1}^{h} (x_i - u)^2 f_i,$$ where n represents the number of data,

h the number of class intervals, and f_i the frequency of the interval containing x_i as the class mark. What would you guess is an equivalent formula using Problem 3.8.7 as a suggestion? It can be

shown to be: $\frac{1}{n} \sum_{i=1}^{h} x_i^2 f_i - \mu^2$. Use this formula to find the standard

deviation of the following data:

x_i (Class mark)	f_i (Frequency)
4	2
6	3
8	3
10	4

3.8.9 If the frequencies in Problem 3.8.8 are all the same, say equal to 4, can you find an easy way to calculate the variance?

3.8.10 Two classes were given an achievement test. One of the classes has a distribution of scores with a standard deviation of 10; the other has a standard deviation of 16. Which of the following statements are true, which are false, and for which is there insufficient information?
(a) Neither of the distributions are skewed.
(b) One distribution has a greater dispersion of scores than the other.
(c) The mean score is smaller for the distribution with the smaller standard deviation.
(d) The mean score is greater for the distribution with the smaller standard deviation.

To determine what per cent of the scores fell within 1σ of the mean, we must examine the class intervals when the data have been put into classes. We want to determine what percent of the scores fell in the $(\mu - 1\sigma, \mu + 1\sigma) = (21.6, 37.4)$ interval. From our frequency table, we know that the class mark of 25 had a frequency of three, 30 had two, and 35 had one.

So, from observation of the class marks, we have found $3 + 2 + 1 = 6$ scores that lie in the interval $(22.5, 32.5)$. However, we must still find the number of scores that fell in the interval on the left, $(21.6, 22.5)$, and in the interval on the right, $(32.5, 37.4)$. The length of the interval on the left is $22.5 - 21.6 = .9$, and so $\frac{.9}{5}$ th of this interval is within 1σ of the mean. Inasmuch as the frequency of the interval is 2, $\frac{.9}{5}(2) = \frac{1.8}{5} = .36$ scores are in the interval $(21.6, 22.5)$. Similarly, the length of the $(32.5, 37.4)$ interval is $37.4 - 32.5 = 4.9$, $\frac{4.9}{5}$ th of this interval also lies within 1σ of the mean. The frequency of the interval is 1, and consequently $\frac{4.9}{5}(1) = .98$ scores are in the interval $(32.5, 37.4)$. Thus we have found the total number of scores that lie within 1σ of the mean: $.36 + 3 + 2 + 1 + .98 = 7.34$. Since $N = 10$, $\frac{7.34}{10} = 73.4$ per cent of the student times fell within 1 standard deviation of the mean. We can conclude that the scores were clustered around the mean.

P R O B L E M S

3.8.11 (a) Using the class marks, determine what per cent of the scores fell with 2σ of the mean; within 1.5σ of the mean.

    (b) Use the raw data to determine the per cent of the scores that fell within 1σ of the mean; 2σ of the mean.

3.8.12 Reorganize your rest pulse data so the class width is 3, and determine what percentage of the scores fell within 1σ and 2σ of the mean.

3.8.13 Follow the same procedure as in Problem 3.8.12 with your jogging pulse data.

3.9 Coding

The raw data we have worked with have not been too cumbersome. However, if the data were quite small (e.g., .000712) or quite large (e.g., 12,273), then determining the mean—although easy in the formula sense—would be difficult numerically. What we can do in such situations is to code the data into a numerically simpler set of data, find the mean, and then convert back to determine the mean of the original data. Although calculators make these computations as easy as any others, the logic of coding, or translating from one set of scores to another, is valuable in itself as it demonstrates how one can create a conversion formula that can be used to relate two sets of data.

Table 3.9.1 is of a frequency distribution of x-scores. We shall use the formula for grouped data to determine the mean. Thus

$$\bar{x} = \sum_{i=1}^{9} \frac{x_i \cdot f_i}{20} = \frac{20 \cdot 2 + 120 \cdot 1 + 220 \cdot 2 + \cdots + 820 \cdot 1}{20} = \frac{8{,}600}{20} = 430$$

Table 3.9.1

x_i-score	Frequency, f_i
20	2
120	1
220	2
320	3
420	2
520	4
620	3
·720	2
820	1
	$\overline{20}$

We begin by coding the data into a new set of data, calculating the mean, and comparing it to the mean above. First, we subtract 20 from each x such that $y_i = x_i - 20$. The frequencies remain the same and the new set of data is: 0, 100, 200, 300, 400, 500, 600, 700, and 800. Thus we have

$$\bar{y} = \sum_{i=1}^{9} \frac{y_i \cdot f_i}{20} = \frac{0 \cdot 2 + 100 \cdot 1 + 200 \cdot 2 + \cdots + 800 \cdot 1}{20} = \frac{8{,}200}{20} = 410$$

It appears that $\bar{y} = \bar{x} - 20$. Does this make sense? Since the frequencies are the same, each of the scores is occurring the same number of times as it did previously. However, since each value is 20 less than it was originally, the new mean should be 20 less. Thus, if we have a set of x-scores as above, we could convert them into y-scores. Then, find $\bar{y}$, add 20 to it, and we obtain $\bar{x}$: $\bar{x} = \bar{y} + 20 = 410 + 20 = 430$.

If we divide each y_i by 100, then $w_i = \dfrac{y_i}{100} = \dfrac{x_1 - 20}{100}$. Thus the w_i-scores are 0, 1, 2, 3, 4, 5, 6, 7, and 8. Then,

$$\bar{w} = \sum_{i=1}^{9} \frac{w_i \cdot f_i}{20} = \frac{0 \cdot 2 + 1 \cdot 1 + 2 \cdot 2 + \cdots + 8 \cdot 1}{20} = \frac{82}{20} = 4.1$$

It appears that $\bar{w} = \frac{1}{100}\bar{y}$. This should happen, since we took $\frac{1}{100}$th of each y_i.

As $\bar{w} = \frac{1}{100}\bar{y}$ and $\bar{y} = \bar{x} - 20$, then

$$\bar{w} = \frac{1}{100}(\bar{x} - 20) = \frac{\bar{x} - 20}{100}$$

Consequently, if we are given a set of data, x_i: 20, 120, 220, 320, 420, 520, 620, 720, and 820, we can code each x_i into a w_i, where w_i represents 0, 1, 2, 3, 4, 5, 6, 7, 8, by setting $w_i = \frac{x_i - 20}{100}$. Then, once we find the mean for the simpler data, $\bar{w}$, we can convert back and find $\bar{x}$. Since $100 \cdot w_i + 20 = x_i$, it follows that $100 \cdot \bar{w} + 20 = \bar{x}$. That is,

$$\bar{w} = 4.1 \qquad \text{and} \qquad \bar{x} = 100 \cdot \bar{w} + 20 = 100 \cdot 4.1 + 20 = 430$$

In general, we can express the relationship between two sets of scores as $x_i = c \cdot w_i + k$, where we multiply a set of data, w_i, by a number c, and then add k to each value; this will cause $\bar{w}$ and $\bar{x}$ to relate as: $\bar{x} = c \cdot \bar{w} + k$. (In our example, $c = 100$ and $k = 20$.)

In an earlier problem, the data shown in Table 3.9.2 were given. Recall we found the mean

$$\bar{x} = \frac{3 \cdot 50 + 5 \cdot 55 + \cdots + 1 \cdot 95}{60} = \frac{4,150}{60} = 69.1\bar{6}$$

Table 3.9.2

Class mark	Frequency
50	3
55	5
60	8
65	12
70	10
75	9
80	6
85	5
90	1
95	1
	60

We shall code x_i so that the arithmetic mean is simpler to obtain. We do this by making some value near the middle, say 70, equal to zero. Subtracting 70 from each score will accomplish this:

$$x_i: 50, 55, 60, 65, 70, 75, 80, 85, 90, 95$$

Thus

$$x_i - 70: -20, -15, -10, -5, 0, 5, 10, 15, 20, 25$$

Now divide by 5: thus,

$$w_i = \frac{x_i - 70}{5} : -4, -3, -2, -1, 0, 1, 2, 3, 4, 5$$

We have created a new set of data, w_i, where $w_i = \frac{1}{5}x_i - 14$; that is, $c = \frac{1}{5}$ and $k = -14$. Now, to find $\bar{w}$ we proceed as follows:

$$\bar{w} = \frac{3(-4) + 5(-3) + 8(-2) + \cdots + 1(5)}{60} = \frac{-10}{60} = \frac{-1}{6}$$

Since $\bar{w} = \frac{1}{5}\bar{x} - 14$, and $\bar{w} = -\frac{1}{6}$, we obtain $\frac{1}{5}\bar{x} - 14 = -\frac{1}{6}$, and hence $\bar{x} = 69.1\bar{6}$, as we found originally.

Coding is a very effective method for simplifying data. For this reason it plays a large role in statistics, as we shall continue to see.

P R O B L E M S

3.9.1 Find the mean for the following data using coding:

x_i	Frequency, f_i
114	1
124	5
134	12
144	16
154	21
164	17
174	16
184	6
194	4
204	1
214	1

3.9.2 Find the mean for the following data using coding:

x_i	Frequency, f_i
61	1
66	5
71	12
76	16
81	21
86	17
91	16
96	6
101	4
106	1
111	1

3.10 Standard Scores

We had said earlier that if two distributions had the same mean value, comparison of their standard-deviation values would clarify the differences between the distributions. If we wanted to compare two scores, one from each distribution, we could determine how far each score was from its respective mean in terms of the standard deviations. By standardizing the scores, that is, by coding the two distributions so that they both have the same mean and standard deviation, a score in one sample can be compared easily to the corresponding score in the other sample. The standardization that would seem to make the most sense is one in which the arithmetic would be quite simple. With this in mind, we shall use the coding techniques developed in Section 3.9 to demonstrate the conversion of any mean and standard deviation to a mean of zero and a standard deviation of 1. In this way, the interval $(\bar{x}-1s, \bar{x}+1s)$ would become $(-1, +1)$, and $(\bar{x}-2s, \bar{x}+2s)$ would become $(-2, +2)$.

Recall that to translate a set of scores, w_i, to a set of scores, x_i, we can use the conversion formulas:

$$x_i = c \cdot w_i + k \qquad \text{or} \qquad \frac{x_i - k}{c} = w_i$$

If $k = \bar{x}$ and $c = s$ (standard deviation), then $w_i = \dfrac{x_i - \bar{x}}{s}$. Because of the special nature of this coding, statisticians refer to the converted scores by a special name: z-scores: $z_i = \dfrac{x_i - \bar{x}}{s}$. Data examined earlier are used in Table 3.10.1 to show the conversion of x- to z-scores. Note in the table that

Table 3.10.1

x_i	$x_i - \bar{x}$	$z_i = \dfrac{x_i - \bar{x}}{s}$
20	−9.5	−1.2
20	−9.5	−1.2
25	−4.5	−.57
25	−4.5	−.57
25	−4.5	−.57
30	.5	.06
30	.5	.06
35	5.5	.69
40	10.5	1.30
45	15.5	2.00

$s = 7.9$ is used in place of $s = 7.8898669$. The mean of the z-scores is found by taking the mean of the scores in the last column:

$$\bar{z} = \frac{2(-1.2)+3(-.57)+2(.06)+.69+1.3+2}{10}$$

$$= \frac{4.11+4.11}{10} = 0$$

To demonstrate that the standard deviation of the z-scores is 1, examine Table 3.10.2. Thus, $s_z = \dfrac{10.028}{10} = 1.0289033$ (approximately 1). The standard deviation would have been exactly 1 if $\sum (z_i - \bar{z})^2 = 10$. This would have been the case if the real value of the standard deviation had been used instead of the approximation, $s = 7.9$.

Table 3.10.2

z_i	$z_i - \bar{z}$	$(z_i - \bar{z})^2$
−1.2	−1.2	1.44
−1.2	−1.2	1.44
−.57	−.57	.3249
−.57	−.57	.3249
−.57	−.57	.3249
.06	.06	.0036
.06	.06	.0036
.69	.69	.4761
1.30	1.30	1.69
2.0	2.0	4.0
		10.0280

Graphically, we have the same histogram as in Figure 3.9.2, but now the x-scores have been standardized to z-scores with a mean of zero and standard deviation of 1 (Figure 3.10.1).

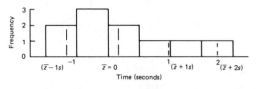

Figure 3.10.1

It is not difficult to convert z-scores to x-scores. For example, if $\bar{x} = 29.5$ and $s = .79$, a z-score of -1.2 can be converted to an x-score of 20:

$$z_i = \frac{x_i - \bar{x}}{s}$$

$$= \frac{x_i - 29.5}{.79}$$

$$-1.2 = \frac{x_i - 29.5}{.79}$$

$$-9.5 = x_i - 29.5$$

$$20 = x_i$$

Standard scores can be applied to examining scores of an individual in different areas of capability. For example, suppose that someone scores 37 on an anxiety scale, 19 on an engineering-as-vocation scale, and 53 on a business-sense scale. These raw scores are not helpful in themselves. However, if they were converted to z-scores, insights can be gained. Suppose that 37 converts to a z-score of .2, 19 to -1.2, and 53 to 1.9. These z-scores suggest that the person has average anxiety, below average in engineering skills, and above average in business aptitude.

We saw earlier that when a set of raw scores, x_i, was multiplied by a constant, c, and k was added, the new set of scores, w_i, is found by the formula $w_i = c \cdot x_i + k$, and the new mean is given by $\bar{w} = c \cdot \bar{x} + k$. To answer the question, "How is the standard deviation, s_x, of this set of x-scores affected?" we take a small sample of scores and find the mean and the standard deviation. Then, we create a new sample by multiplying and adding a constant to each score.

Suppose that we have the raw data $x_1 = 3$, $x_2 = 4$, and $x_3 = 8$. Then:

1. $\bar{x} = \dfrac{3+4+8}{3} = \dfrac{15}{3} = 5.$

2. $s_x = \sqrt{\dfrac{1}{3-1} \sum\limits_{i=1}^{3} (x_i - \bar{x})^2} = \sqrt{\tfrac{1}{2}[(3-5)^2 + (4-5)^2 + (8-5)^2]}$

$= \sqrt{\tfrac{1}{2}(14)} = \sqrt{7} = 2.65$

3. Since $w_i = c \cdot x + k$, we let $c = 1$ and $k = 2$, and $w_i = x_i + 2$. That is, we are not changing the x_i by multiplying but only adding 2 to each x_i. Thus, we shift each x_i 2 units to the right. We obtain that $\bar{w} = 1 \cdot \bar{x} + 2 = 5 + 2 = 7.$ This can be seen as $w_1 = 5$, $w_2 = 6$, $w_3 = 10$, so $\bar{w} = \dfrac{5+6+10}{3} = \dfrac{21}{3} = 7.$

4. $s_w = \sqrt{\dfrac{1}{3-1}[(5-7)^2 + (6-7)^2 + (10-7)^2]} = \sqrt{\tfrac{1}{2}(14)} = \sqrt{7} = 2.65.$ That

is, $s_w = s_x$. Adding a constant k shifts all the scores k units. It should be clear that the deviation from the mean does not change, because the mean has also shifted k units.

5. Let $c = 2$ and $k = 4$; then, if we double each score and add 4 to each score, $w_i = 2x_i + 4$, so $w_i = 10$, 12, 20. From this, $\bar{w} = \dfrac{10 + 12 + 20}{3} = \dfrac{42}{3} = 14$, and $\bar{w} = 2 \cdot \bar{x} + 4$, as we would expect.

6. $S_w = \sqrt{\frac{1}{2}[(10-14)^2 + (12-14)^2 + (20-14)^2]} = \sqrt{\frac{1}{2}(56)} = \sqrt{28} = \sqrt{4} \cdot \sqrt{7} = 2\sqrt{7} = 5.30$. Since $s_w = 2 \cdot s_x$, we can conclude that doubling each score doubles the dispersion. The original data of 3, 4, and 8 show that 1 unit is between the first two scores $(4-3)$ and 4 units are between the last two scores $(8-4)$. Now, as we doubled the scores, the data are 10, 12, 20—there is 2 units between the first two scores and 8 units between the last two. The dispersion is doubled when the data are doubled. Adding the constant had no effect on the standard deviation, as we saw in points 2 and 4 above.

In general, each x_i can be converted to a w_i, and conversely, each w_i can be converted to an x_i. We have determined that

$$w_i = c \cdot x_i + k \qquad \text{or} \qquad \frac{w_i - k}{c} = x_i$$

$$\bar{w} = c \cdot \bar{x} + k \qquad \qquad \frac{\bar{w} - k}{c} = \bar{x}$$

$$s_w = c \cdot s_x \qquad \qquad \frac{1}{c} \cdot s_w = s_x$$

PROBLEMS

3.10.1 (a) Find the standard deviation of the population data given in Problem 3.9.1 in its coded form and convert to the standard deviation of the raw scores.

(b) Convert the population data to z-scores.

(c) Convert the interval $(-2, +2)$ to the original interval.

3.10.2 If $\bar{x} = 75$ and $s_x = 10$, find the corresponding z-score when $x = 82$.

3.10.3 If $s = 10$, and an x-score of 90 is converted to a $z = 2.5$, what is the value of $\bar{x}$?

3.10.4 If $c = 10$ and $k = 4$, find:

(a) x_i if $w_i = 1{,}604$.

(b) w_i if $x_i = 1{,}600$.

3.11 Summary

In this chapter we have introduced measures of central tendency and variability. The three averages—mode, median, and mean—play important roles in examining data. The mode is the score that occurs the most often in a set of scores and requires the data to be organized. Its major weakness is that it does not tell us anything regarding the rest of the data and so has a minimal role in statistics. The median is the score that falls above 50 per cent of the scores and below 50 per cent of the scores. To determine the median, the data must be organized in ascending order. In the area of inferential statistics, where we make judgments about the populations from the samples we examine, the median has a minor role. The average that has widest usage in inferential statistics is the arithmetic mean. The mean does not require the organizing or ordering of data. It tells us the center of balance of the distribution.

Although the three measures of central tendency do give us insight into the collected data, the need for a statistic that considers dispersion or variability about the average is crucial to an understanding of the distribution of the data. Four measures of dispersion were considered: range, percentile, variance, and standard deviation. The range is a statistic that measures the difference between the highest and the lowest scores. The range has limited usage inasmuch as it does not tell us anything about the shape. When the median was used as the average, the 25th and 75th percentiles were used as measures of variability. The 25th percentile (Q_1) and the 75th percentile (Q_3) represent the numbers that 25 per cent of the scores are below and above, respectively. A major application of the median and the percentiles is in interpreting standardized test scores. For example, a score that falls in the 83rd percentile, $P_{.83}$, infers that 17 per cent of the scores were higher. The variance is a measure of dispersion about the mean. One drawback of the variance is the fact that it is a measure using squared units and is not easily related to the histogram. The standard deviation, the square root of the variance, is the measure of dispersion about the mean. By knowing the standard deviation of a set of scores about the mean, we can get a fairly good picture of the distribution of scores.

3.12 Review of Formulas

   I. Measures of central tendency
      A. For ungrouped data
         1. Sample mean for n-scores: $x_1, x_2, \ldots, x_n$:

$$\bar{x} = \sum_{i=1}^{n} \frac{x_i}{n} = \frac{x_1 + x_2 + \cdots + x_n}{n}$$

B. For grouped data
   1. Sample mean for n-scores that fall into k groups, where f_i is the frequency of the first group, f_2 the frequency of the second group, and f_k the frequency of the kth group:

$$\bar{x} = \frac{\sum\limits_{i=1}^{k} x_i \cdot f_i}{n} = \frac{x_1 f_1 + x_2 f_2 + \cdots + x_k f_k}{n}$$

and

$$n = \sum_{i=1}^{k} f_i$$

   2. Population mean

$$\mu_x = \frac{\sum\limits_{i=1}^{k} x_i \cdot f_i}{N} = \frac{x_1 f_1 + \cdots + x_k f_k}{N}$$

and

$$N = \sum_{i=1}^{k} f_i$$

   3. Median

$$m = l + \frac{f}{N}(w)$$

   where l is the lower bound of the class interval containing the median, f the frequency needed to reach the median, N the number of scores in the interval containing the median, and w the width of the interval.

II. Measures of variability
   A. For ungrouped data
      1. Sample standard deviation for n pieces of data: $x_1, x_2, \ldots, x_n$ and mean x:

$$S_x = \sqrt{\frac{1}{n-1} \sum_{i=1}^{n} (x_i - \bar{x})^2}$$

      2. Population standard deviation for N pieces of data: $x_1, x_2, \ldots, x_n$ and mean μ_x:

$$\sigma_x = \sqrt{\frac{1}{N} \sum_{i=1}^{N} (x_i - \mu_x)^2} = \sqrt{\frac{1}{N} \sum_{i=1}^{N} x_i^2 - (\mu_x)^2}$$

B. For grouped data
  1. Sample standard deviation for k groups and n pieces of data:

$$s_x = \sqrt{\frac{1}{n-1} \sum_{i=1}^{k} (x_i - \bar{x})^2 \cdot f_i}$$

  2. Population standard deviation:

$$\sigma_x = \sqrt{\frac{1}{N} \sum_{i=1}^{k} (x_i - \mu_x)^2 \cdot f_i}$$

  3. First and third quartiles:

$$Q_1 = l_{25} + \frac{f}{n}(w) \qquad \text{and} \qquad Q_3 = l_{75} + \frac{f}{N}(w)$$

where l_{25} and l_{75} are the lower bounds of the class interval containing the 25th and 75th percentile.

3.13 Problem Set

3.13.1 Find the mean, median, and mode for the following scores: 10, 15, 4, 9, 4, 5, 2, 12, 3, 4, 3, 5, and 8.

3.13.2 Using the bus-stop data on p. 29, which of the measures of central tendency would you choose to determine if the stop is necessary? Why?

3.13.3 Find the median (Q_2), Q_1, Q_3, and $P_{.95}$, for the following distribution of scores:

Class interval	Frequency
7.5–10.5	2
10.5–13.5	3
13.5–16.5	5
16.5–19.5	7
19.5–22.5	4
22.5–25.5	3
25.5–28.5	1
	$\overline{25}$

3.13.4 Refer to Table 2.3.2 and determine:
  (a) What score fell above and below 50 per cent of the scores?
  (b) What is the score that fell above 25 per cent of the scores?
  (c) What is the score that fell above 80 per cent of the scores?

3.13.5 Using class marks for the sample data in Problem 3.13.3, determine:
   (a) The mean.
   (b) The variance.
   (c) The standard deviation.
   (d) The per cent of scores that fell within 1 standard deviation of the mean.

3.13.6 In a class of 40 students, the mean was 76 and the standard deviation was 4. If the student's grade was 2 standard deviations above the mean, what was the grade?

3.13.7 Given the data 4, 7, 11, 15, and 18:
   (a) What is the new mean when 6 is added to each score?
   (b) What is the new mean when each score is multiplied by 100?
   (c) What happens to the standard deviation when 6 is added to each score?
   (d) What happens to the standard deviation of a set of scores when each score is multiplied by 100?

3.13.8 (a) Find the mean and standard deviation, by coding, of the following population data: 1,015, 1,020, 1,020, 1,025, 1,030, 1,035, and 1,040.
   (b) What per cent of the scores fell within 1 standard deviation of the mean?

3.13.9 (a) Convert the following sample scores to z-scores: 61, 73, 80, 84, 92, and 96.
   (b) Show that the mean of the z-scores is zero.
   (c) What per cent of the z-scores fell in $(-1, 1)$ interval?
   (d) What per cent of the x-scores fell within 1 standard deviation of the mean?

3.13.10 Transform the following set of population data—2, 8, 8, 10, 10, 12, 14, 16—into a set of scores with a mean of 50 and a standard deviation of 10.

3.13.11 (For algebra buffs.) Prove that

$$\frac{1}{n} \sum_{i=1}^{n} (x_i - \bar{x})^2 = \frac{1}{n} \sum_{i=1}^{n} x_i^2 - \bar{x}^2$$

3.13.12 Refer to those examples in which you have determined the 25th, 50th, and 75th percentiles. In these examples compare $Q_3 - Q_2$ with $Q_2 - Q_1$. Can you make any conjectures as to the relationship of $Q_3 - Q_2$ with $Q_2 - Q_1$ and the skewness of the distributions?

3.13.13 Why do you imagine we did not mention variation about the mode?

3.13.14 Suppose that you have the chance to take any of three jobs for the summer. The mean wage and standard deviation for each of the three companies is given in the following table. Which company pays the highest overall wages (consider three standard deviations to the right of the mean)? Which company pays the most consistently? Which company has the greatest wage differences?

	Mean wage	Standard deviation
A	$2.05	.58
B	2.75	.75
C	2.35	.95

3.13.15 Find the thickness of a page of this book. How does this problem relate to this book? Is it possible that your answer does not in fact represent the thickness of any page in this book? Explain.

3.13.16 Find the average and standard deviation of the heights of the students in your class. Graph the data in the form of a histogram.

3.13.17 Repeat Problem 3.13.16, but this time separate the females from the males and create two graphs.

3.13.18 Standardize the three distributions—the total class, the males, and the females. What conclusions can you draw?

3.13.19 Find the length between your nose to your fingers of an extended arm. Find the average for your class. Find the mean for those students who are around 6 feet tall. Is it close to a yard?

3.13.20 Suppose that you were president of company A. Which average would you use to prove that company A produces more than company B on the average?

$$A: \ 478, 503, 505, 513, 516$$
$$B: \ 511, 486, 512, 486, 510$$

Now suppose you were president of company B—which average would you choose to prove that company B produces more than company A on the average?

3.13.21 Which collection of data—A or B—would you use to find the four-year over-all batting average of player X:

Table A			Table B		
Year	Batting average		Year	Batting average	Number of times at bat
1	.286		1	.286	364
2	.275		2	.275	298
3	.292		3	.292	435
4	.278		4	.278	428

Why? Find the over-all average.

Chapter 4
Probability

4.1 Introduction

The descriptive statistics that we examined in Chapters 2 and 3 provide a basis for our future work in inferential statistics. One area that we need to consider before we make the leap from describing quantitative situations to predicting them is the concept of chance. Almost all human experience contains an element of chance, including this present experience. In situations that are important to us, we try to minimize the chance of making a poor judgment. The chance of success depends upon getting objective and relevant information. The better the information, the better the chance of success and the smaller the chance of failure. When we understand a situation to the extent that we know which information is useful and which is not, we are a long way toward solving the problem.

The value of relevant information can be seen in the following statement by Werner Von Braun, who was a leading rocket expert, as he compared the testing of rockets in 1961 and in 1967:

> Then it was a kind of trial and error method. You designed these things but didn't really know what they would do until you pushed the button. Through the use of computer analysis, vibration tests, simulations of the space environment and ground firings, it has been possible to weed out or redesign potentially troublesome parts in the rocket. We have really tortured all the systems components against all conceivable environmental conditions the rocket might see during launch or flight. And we have an automatic electronic checkout system that monitors thousands of points on the rocket right up to the moment of lift-off so that the probability of launching a sick system is quite remote.*

In order to evaluate situations realistically, we need to assign reasonable probabilities to events. The Weather Bureau, like the rest of us, uses past experience to predict the chance of future occurrences. Sometimes the number of experiences that we use to make a judgment is too small—the information that we have collected is only from a particular moment; a few days; a few people. In these cases the "past experience" may lead us to incorrect decisions. In this chapter we shall explore the basic ideas of probability and some of its applications.

---

* *The New York Times*, November 5, 1967.

4.2 Chance Based on Experience

The boundaries of chance extend from the realm of certainty to the realm of impossibility. Sometimes we use experience to determine chance and sometimes we use logic. A 20 per cent chance of rain means that 20 times out of 100, experience has shown that, with certain conditions existing, it will rain. That is, Chance(rain) = $\frac{20}{100}$. The numerator represents the number of favorable outcomes that are favoring the occurrence of the event, and the denominator represents the total number of possible outcomes. If the numerator were 0, the chance of rain would be $\frac{0}{100} = 0$; and if the numerator were 100, the chance of rain would be $\frac{100}{100} = 1$. So chance lies between impossibility (zero) and certainty (1), see Figure 4.2.1.

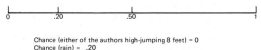

Chance (either of the authors high-jumping 8 feet) = 0
Chance (rain) =   .20
Chance (in the flip of a fair coin, heads landing face up) = .50
Chance (neither of the authors high-jumping 8 feet) = 1

Figure 4.2.1

Chance(either of the authors could high-jump 8 feet) = 0
Chance(rain) = .20
Chance(in the flip of a coin heads landing face up) = .50
Chance(sum of the authors' jumps would not be over 8 feet) = 1

We can determine the chance of a coin landing heads as being .50, as there are two possible outcomes:

$$\text{Chance(heads)} = \frac{1 \text{ (numbers of favorable outcomes)}}{2 \text{ (number of possible outcomes)}} = 50 \text{ per cent}$$

Suppose that we flipped a coin 1,000 times and found that heads landed face up 489 times; then Chance(heads) = $\frac{489}{1,000}$ = 48.9 per cent. Which is correct— 48.9 per cent or 50 per cent? Our reason leads us to 50 per cent and our experience to 48.9 per cent. If logic and experience agree, fine. If they don't, either our logic is faulty or our assumptions are not valid; for example, the coin is weighted and is not fair. In this situation, since .50 is very close to .489, we can believe that the coin is fair. We do not expect a coin to fall exactly 50 per cent heads but rather around 50 per cent. Suppose that we had obtained 42 per cent heads, would you still believe the coin is fair? Suppose 39 per cent? The question of how much deviation from what we expect is still acceptable will be considered in Chapter 6.

If we flipped a coin 5 times annd obtained 1 head, we would say based on experience that Chance(heads) $= \frac{1}{5} = 20$ per cent. Perhaps it is for this coin (if it is loaded), or perhaps our sample is so small (only 5) that if we tossed the coin 95 more times for a total of 100 tosses, we might find 47 heads. Whenever our logic is contradicted by our "common sense" or experience, the situation must be reexamined. Chance based upon experience is referred to as *relative frequency*. Like experience, relative frequency is always changing. For example, if we flipped a coin 6 times and obtained 3 heads, we would say that the Chance(heads) $= \frac{3}{6} = .5000$. Now on the seventh toss, if the coin falls heads, the Chance(heads) $= \frac{4}{7} = .5714$; or if it falls tails, Chance(heads) $= \frac{3}{7} = .4286$. The change in our relative frequency reflects our changing knowledge. Over the long run, the relative frequency will change less and less, and the fraction that the relative frequency is approaching is called the *probability*. Thus, if we flipped a coin, 1,000 times and found 497 heads, then $\frac{497}{1,000} = .4970$ is the relative frequency of obtaining heads. If the next flip produced a heads, we would have $\frac{498}{1,001} = .4975$, which differs very little from .4970, or if it were tails we would have $\frac{497}{1,001} = .4965$, which also differs by very little. In a sense, the relative frequency "searches out" the true probability. If logic does not dictate a particular probability value, we can use the concept of relative frequency to find it.

We change our beliefs and attitudes as we learn and experience. The more information we get, the more we must make adjustments, and the clearer the truth becomes. Consider the following situation, where relative frequency works like a "divining rod" as it searches out the true probability measure. In the throw of a single dice (a die), the six possible outcomes are 1, 2, 3, 4, 5, and 6. We would expect that the chance of obtaining a "4" is equal to any of the other outcomes and would be one chance out of six, $\frac{1}{6}$, as the die is assumed to be weighted equally. Table 4.2.1 presents a frequency distribution of 996 throws of this die and the expected distribution. We see that the outcomes actually obtained are not at all what

Table 4.2.1. Expected and Observed Frequency Distribution
of Throwing a Die

Number	Expected outcomes	Observed outcomes
1	166	165
2	166	159
3	166	3
4	166	350
5	166	161
6	166	158
	996	996

we would expect if the die were fair. We should have expected about $\frac{1}{6}$ of the outcomes, approximately 17 per cent, to have been a "4," whereas we observed $\frac{350}{996}$ of the outcomes, approximately 35 per cent. Thus we should tend to reject our assumption that the coin is fair.

In fields such as insurance, advertising, and medicine, we frequently have no logical scheme to determine what probability values to assign, so that we must depend upon experience and relative frequency to determine the probability of these events. This concept will be explored in greater detail when we consider inferential statistics.

P R O B L E M S

4.2.1 Toss a tack 10 times and find the chance that the tack lands on its side, and the chance of landing point up. Combine your results (if the same type of tack was used) with those of four other students (total of 50 tosses) and determine the relative frequency of landing point up; then with five more students (total of 100 tosses); and finally, with the entire class. What can you say about the four relative frequency measures you found (10, 50, 100, and class total)? Compare your "50 toss" results with others who used a different sample. What can you say about these "50 toss" averages?

4.2.2 *Class Project*: Reexamine your rest pulse data and find the chance that a score picked at random is:
(a) Equal to 50, 55, 60, 65, 70, 75, and 80.
(b) Less than 75, at least 75.
(c) Between 50 and 60, 55 and 65, 65 and 75, 60 and 70, and 70 and 80.

4.2.3 Respond to Problem 4.2.2 using your jogging pulse data.

4.2.4 Find the chance that students in your class will watch television on an "average" night between 0 and 1 hours, 1 and 2 hours, 2 and 3 hours, 3 and 4 hours, and 4 and 5 hours. What, if any, conclusions can be drawn? Draw a histogram.

4.2.5 *Class Project*: In your class, what is the chance that if a student were drawn randomly from the class register, that student would have 0 siblings, 1 sibling, 2 siblings, 3 siblings, 4 siblings, or 5 siblings.

4.2.6 From the data you collected from Problem 4.2.5 determine the chance that a student in your class has:
(a) At most 1 sibling, at least 2 siblings.
(b) At least 3 siblings, less than 3 siblings.
(c) More than 2 siblings.

(d) Between 0 and 1 sibling, 0 and 2 siblings, between 3 and 5 siblings.

4.3 Chance Based on Logic

We saw that chance determined by experience is known as relative frequency. If we knew beforehand (as with a fair coin or fair die) all the possible outcomes and that each outcome has an equal chance, it would not be necessary to choose a sample, organize the data, and determine the relative frequency. Instead we could use our logic to determine the probability of certain outcomes, that is, chance based on logic is referred to as probability. One way to test if our logic is correct in a chance situation is to compare our probability value with the relative frequency value we obtain from experience. Provided that our assumptions are correct, over the long run, that is, as the sample size increases, the relative frequency value will be very close to the probability value that we determined logically. For example, in a deck of 52 playing cards, there are 13 hearts, 13 spades, 13 diamonds, and 13 clubs. Using logic, we would determine that the Prob(picking a heart in one try) $= \frac{13}{52} = .25$. If we drew a card, replaced it, shuffled, and picked again, repeating the process over and over, the relative frequency of hearts to total cards picked would approach 25 per cent.

If we flipped three coins 10 times and recorded the number of heads on each toss of three coins, we might obtain the frequency histogram of Figure 4.3.1. The relative frequencies are shown in Table 4.3.1. Would this distribution occur again if we did the experiment 10 more times? Maybe

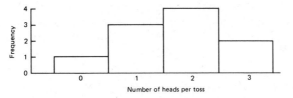

Figure 4.3.1 Histogram of Flipping Three Coins 10 Times

Table 4.3.1. Observed Relative Frequencies of Flipping Three Coins 10 Times

Number of heads	Relative frequency
0	$\frac{1}{10}$ (.10)
1	$\frac{3}{10}$ (.30)
2	$\frac{4}{10}$ (.40)
3	$\frac{2}{10}$ (.20)

yes and maybe no. However, we do expect that in the long run the relative frequency would move closer and closer to the probability value.

To find the probability of this event, we shall first determine the collection of possible outcomes and the probability associated with each outcome. One way to do this is to create a *tree diagram*, which is a diagram made up of branches showing the outcomes on each trial of the event. In our case we wish to find all the possible outcomes for flipping a coin three times (which is equivalent to flipping three coins once). On each toss of a coin, the possibilities are heads or tails. Hence our tree diagram would appear as shown in Figure 4.3.2. We repeat heads and tails twice on the

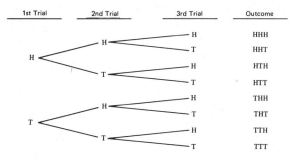

Figure 4.3.2 Tree Diagram for Flipping Three Coins

second trial and four times on the third trial because each flip of the coin produces a head or a tail, and either heads or tails can be associated with the outcomes of the previous trial. The possible outcomes for flipping three coins are listed in the last column of Figure 4.3.2. Since $P(H) = P(T)$, we find that $\frac{1}{8}$ of the time we should obtain three heads: *HHH*; $\frac{3}{8}$ of the time, two heads and a tail: *HHT, HTH, THH*; $\frac{3}{8}$ of the time, two tails and a head: *HTT, THT, TTH*; and $\frac{1}{8}$ of the time, three tails: *TTT*. The theoretical (derived by logic) probability distribution is shown in Table 4.3.2. Over the long run, we would expect the observed relative frequency distribution to be quite close to the theoretical probability distribution.

Table 4.3.2. Theoretical Probability Distribution for Flipping Three Coins

Number of heads	Number of tails	Probability
0	3	$\frac{1}{8}$ (.125)
1	2	$\frac{3}{8}$ (.375)
2	1	$\frac{3}{8}$ (.375)
3	0	$\frac{1}{8}$ (.125)

P R O B L E M S

4.3.1 Construct a histogram for the probability column of Table 4.3.2 and compare it with Figure 4.3.1.

4.3.2 Flip three coins 10 times and combine your results with those in Table 4.3.1. That is, you now have $n = 20$ outcomes. Is your new relative frequency distribution closer to that of Table 4.3.2?

4.3.3 (a) Create a tree diagram to find the probability that in the births of three children there are: all girls, two girls and one boy, one girl and two boys, and all boys.

(b) Use a tree diagram to find the possible outcomes and associated probabilities when there are four births.

(c) Find the probabilities from part (b) of having at least two (two or more) girls; between one and three boys (inclusive); and at least one boy.

We said earlier that in many fields, such as medicine and advertising, responses and percentages of responses are not known before the activity is performed. If we think we have a good theory to explain what will happen, we can create a mathematical model—a theoretical picture of which we think will occur—and test it against the real-world outcomes. If we are correct, the outcomes can be predicted by the model; if not, our assumptions are incorrect, and we must create a new model to explain outcomes.

Psychology is one of the areas in which mathematical models are tested against reality in order to understand how human being act. We shall consider a psychology experiment and see if we can create some mathematical models that will predict behavior for this situation. The situation we shall examine is known as "Response to Stimulus Compound," that is, response made to more than one stimulus.* The stimuli consist of 24 light bulbs: three rows, each containing 8 lights. The first row will be referred to as S_1 (stimulus 1); the second row, S_2; and the third row, S_3. The subjects in the experiment were instructed to flip a switch up if lights went on in S_1, down if lights went on in S_2. During the training session, no lights in S_3 were put on or even discussed. The subjects were told to respond as they saw fit. For instance, if 6 lights went on in S_1 and 2 lights went on in S_2, we would guess that the subject would flip the switch up, since more lights went on in the top row.

Table 4.3.3 contains the data from eight test patterns and the average proportions of up responses observed. For example, the first test had all 8

* E. R. Hilgard and G. H. Bower, *Theories of Learning*, 3rd ed. (New York: Appleton–Century–Crofts, 1966).

lights on in both S_1 and S_2, and none in S_3; 54 per cent of the time the
switch was turned up when the pattern was 8, 8, 0.

Table 4.3.3. Responses to Eight Test Patterns

Test pattern	Number of lights on			Observed $P(up)$
	S_1	S_2	S_3	
1	8	8	0	.54
2	8	4	0	.79
3	8	2	0	.81
4	4	2	0	.63
5	8	4	8	.62
6	8	2	8	.67
7	4	2	8	.54
8	8	8	8	.54

Look over the response data and see if you find any interesting results.
Some that come to our attention are

1. In test 1, although $n(S_1) = n(S_2) = 8$, the response was not exactly 50
   per cent for each. Does this suggest a reason why candidates running
   for office always try to have their names on the first line of the ballot?
   (It would be interesting to design an experiment that would test
   whether first-row "candidates" do better—without the subjects
   knowing anything about the candidates.)
2. In tests 3 and 6, while $n(S_1) = 8$, and $n(S_2) = 2$, $P(up) = .81$ and .67,
   respectively. And too, in tests 2 and 5, $n(S_1) = 8$, and $n(S_2) = 4$, while
   $P(up) = .79$ and .62, respectively. Is S_3 favoring "down"? Look at S_3
   in each of these four test patterns.

When we create a model, it is based on certain assumptions. If the
assumptions are correct, the model will be a good predictor of the real
outcomes. We shall compare the predictions made by three models to the
data in Table 4.3.3.

MODEL 1

Assumptions
1. If there are some more lights on in S_1 than in S_2, the subject would
   throw the switch up all the time; that is, model 1 predicts that if
   $n(S_1) > n(S_2)$, then P(switch up) = 1.
2. If $n(S_1) = n(S_2)$, then P(switch up) = P(switch down) = .50.
3. S_3 has no effect because it was not discussed.

We can compare the observed values to the predicted values using this model's assumptions (Table 4.3.4). It is evident that this "majority-rule" model is not appropriate.

Table 4.3.4. Model 1

Test pattern	Observed	Predicted
1	.54	.50
2	.79	1.00
3	.81	1.00
4	.63	1.00
5	.62	1.00
6	.67	1.00
7	.54	1.00
8	.54	.50

MODEL 2

Assumptions

1. S_3 does not affect the choice.
2. P(switch up) is determined by the relative frequency of lights on in S_1 to the total lights on in S_1 and S_2; that is, P(switch up) =

$$\frac{n(S_1)}{n(S_1)+n(S_2)} = \text{the ratio of the number of favorable outcomes to the}$$

total number of outcomes.

The comparison of model 2 predictions and the observed percentages are found in Table 4.3.5. This model supplies a pretty good "fit" for test patterns 1, 3, 4, 5, and 8.

Table 4.3.5. Model 2

Test pattern	Observed	Predicted
1	.54	.50
2	.79	.67
3	.81	.80
4	.63	.67
5	.62	.67
6	.67	.80
7	.54	.67
8	.54	.50

MODEL 3

Assumptions

1. The relative frequency model used in model 2 is acceptable, but
2. S_3 should be considered. We shall add $n(S_3)$ to the total outcomes and assign $\frac{1}{2}$ of $n(S_3)$ to each direction of up and down, since S_3 was not associated with S_1 or S_2; that is, $P(\text{up}) = \dfrac{n(S_1) + \frac{1}{2}(S_3)}{n(S_1) + n(S_2) + n(S_3)}$.

The comparison of model 3 predictions and the observed percentages are given in Table 4.3.6. We note that in test patterns 1 to 4 and 8 that model 2 and 3 produced the same values, but that in test patterns 5 to 7, model 3 is a much better predictor. It appears that model 3 supplies the better approximation of this experiment. However, there are many other test patterns that could be created. We can only conclude that up to this time, the logic of model 3 creates the best fit for the data.

Table 4.3.6. Model 3

Test pattern	Observed	Predicted
1	.54	.50
2	.79	.67
3	.81	.80
4	.63	.67
5	.62	.60
6	.67	.67
7	.54	.57
8	.54	.50

P R O B L E M

4.3.4 A ninth test pattern, where $S_1 = 8$, $S_2 = 0$, and $S_3 = 8$, resulted in $P(\text{up}) = .73$. Determine the predicted percentages for each of the three models.

When we find the probability of some event happening, we can easily determine the probability of that event not happening. For example, the first test pattern resulted in 54 per cent of the subjects pushing the switch up; that is, $P(\text{up}) = .54$. If 54 per cent pushed the switch up, then 46 per cent, $(100 - 54)$ per cent must have pushed the switch down. That is, $P(\text{up}) + P(\text{down}) = 1$, certainty. Since the switch could go only up or down, $P(\text{up or down}) = P(\text{up or not-up}) = 1$. In general, if E is an event, then not-E, the complement of E, is written as E'; and $P(E \text{ or } E') =$

$P(E) + P(E') = 1$. Since

$$P(E) = \frac{\text{number of simple outcomes favorable to } E}{\text{total number possible}}$$

then

$$P(E)' = \frac{\text{number of simple outcomes not favorable to } E}{\text{total number possible}}$$

as nothing else can occur other than E or E'. For example, since the chance of having three girls in the birth of three children, $P(\text{girls} = 3) = \frac{1}{8}$, the chance of not having three girls (this does not mean that all three are boys), $P(G \neq 3) = \frac{7}{8}$, as $P(G = 3 \text{ or } G \neq 3) = 1$. That is, $P(G = 3 \text{ or } G \neq 3) = P(G = 3) + P(G \neq 3) = \frac{1}{8} + \frac{7}{8} = 1$.

The question that we must consider is: If any two outcomes, A and B, of a particular experiment occur, does $P(A \text{ or } B) = P(A) + P(B)$? If we look at Table 4.3.2, we find that $P(H = 1 \text{ or } H = 2) = P(H = 1) + P(H = 2) = \frac{3}{8} + \frac{3}{8} = \frac{3}{4}$. It appears that the rule works. But we should not create a generalization based on only one case. We can use Table 4.3.2 again to determine $P(H \geq 1 \text{ or } H = 2)$, that is, the chance that the number of heads is at least 1 (1 or more) or is 2. Does $P(H \geq 1 \text{ or } H = 2) = P(H \geq 1) + P(H = 2)$? $P(H \geq 1) = \frac{7}{8}$ and $P(H = 2) = \frac{3}{8}$, so we have $\frac{7}{8} + \frac{3}{8} = \frac{10}{8} > 1$. But we know that the probability of any event in a given experiment cannot be greater than 1. Since $P(E \text{ or } E') = 1$, and either E or E' can happen and nothing else, then something is wrong when we get a probability value greater than 1, certainty.

Recall that $P(H \geq 1) = P(H = 1 \text{ or } H = 2 \text{ or } H = 3)$. Therefore, when we find $P(H \geq 1 \text{ or } H = 2)$, we have found $P(H = 1) + P(H = 2) + P(H = 3) + P(H = 2)$; we have counted "$P(H = 2)$" twice. Because it may only be counted once, as is true for any event, we would determine that $P(H \geq 1 \text{ or } H = 2) = P(H \geq 1) + (H = 2) - P(H = 2) = \frac{7}{8} + \frac{3}{8} - \frac{3}{8} = \frac{7}{8}$. That is, we have eliminated the extra event that was counted in both events. In general, for any events A and B

$$P(A \text{ or } B) = P(A) + P(B) - P \text{ (any simple outcome common to both } A \text{ and } B)$$

What about our old rule: $P(A \text{ or } B) = P(A) + P(B)$—when does it work? Suppose that A and B have no simple outcome in common; then P(events common to A and B) = 0. That is, $P(A \text{ or } B) = P(A) + P(B)$ is appropriate when A and B have no events in common. So our new generalization includes our original rule.

We have determined three rules to use in the logic of probability:
1. $0 \leq P(A) \leq 1$; that is, the probability of any event is at most 1 and at least zero.

2. $P(A \text{ or } A') = P(A) + P(A') = 1$.
3. $P(A \text{ or } B) = P(A) + P(B) - P(\text{simple outcomes common to } A \text{ and } B)$.

PROBLEMS

4.3.5 Can events A and A' have any events in common?

4.3.6 When two dice are thrown, the sum of the numbers on the faces of the dice can be any number from 2 $(1 + 1)$ up to 12 $(6 + 6)$, as each die contains the numbers from 1 to 6. The possible sums are 2, 3, 4, 5, 6, 7, 8, 9, 10, 11, and 12. Choosing one of the possible sums, find:
   (a) $P(\text{even number})$.
   (b) $P(\text{odd number})$.
   (c) $P(\text{even or odd number})$.
   (d) $P(\text{number greater than 7})$.
   (e) $P(\text{odd or divisible by 4})$.
   (f) $P(\text{even or divisible by 3})$.

4.3.7 An English teacher was interested in her students' reactions in comparing two articles on the same subject. She asked the class to glance at both articles and to choose one to read first and to read the other if it interested them. The class, knowing their teacher's interest in statistical analysis, presented her with the following data: 91 per cent of the total group read at least one of the articles; 75 per cent of the group read article A; and 66 per cent of the group read both articles. What per cent read article B?

4.3.8 What type of distribution is represented by the expected-outcomes column in Table 4.2.1?

4.4 Multiplication Principle

If you had to choose an outfit consisting of one of each of the following— 5 sweaters, 7 shirts, 6 pairs of pants, 4 pairs of shoes, and 4 jackets—there would be 3,360 possible outfits from which to choose. You might not want to wear some of the outfits, but there are surely enough to choose from. The question we want to answer is: How is it that there are 3,360 possible outfits (enough outfits for 9 years)?

We could create a tree diagram to determine all the possibilities. However, this would require the drawing of 3,360 branches on the diagram. Instead, we will look at a shortened version of the problem and try to find a pattern that will eliminate much of the busy work.

If we considered the outfits that could be created from 2 jackets, 3 shirts, and 2 pairs of pants, a tree diagram would show that there are 12 possible

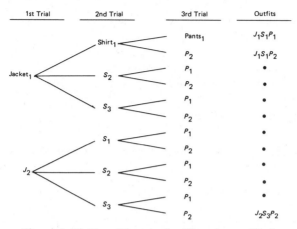

1st Trial	2nd Trial	3rd Trial	Outfits

Figure 4.4.1 Tree Diagram for Choosing an Outfit

outfits (Figure 4.4.1). If there were just 2 jackets and 3 shirts, there would be 6 possible outfits containing both. If you examine the number of choices and the corresponding number of events, you observe that since $2 \times 3 = 6$ (6 outfits containing a jacket and a shirt) and $2 \times 3 \times 2 = 12$ (12 outfits containing a jacket, a shirt, and a pair of pants), we have, in general, a *multiplication principle for counting*: If there are n ways of doing a first event and m ways of doing a second event, there are $n \times m$ ways of doing both events. This principle can be extended for any number of events. For example, using the principle on our "outfit" problem, we find that there are $5 \times 7 \times 6 \times 4 = 3,360$ five-piece outfits from which to choose. Similarly, a tree diagram showing the outcomes of flipping five coins would have $2 \times 2 \times 2 \times 2 \times 2 = 32$ branches listing each of the 32 quintuples of outcomes.

Multiplication is used when more than one outcome *must* occur. In picking a five-piece outfit, for example, we had to pick a sweater *and* a jacket *and* a pair of pants. In contrast, addition is used when at least one or more outcomes must occur. For example, in how many ways can you contact a friend? You might write *or* call *or* visit. In this situation, we would not use *and*, since either writing *or* calling *or* visiting would satisfy the event of getting in contact with your friend. If there were two ways of writing (letter or telegram), one way of calling (phone), and three ways of visiting (plane, car, bicycle), there would be $2 + 1 + 3 = 6$ ways of getting in touch with your friend. That is, any of these six ways would satisfy the situation. However, the number of ways of choosing jacket *and* sweater *and* pants is $5 \times 7 \times 6 = 210$; one of each is required.

The multiplication principle is also helpful in finding the probability of both A and B occurring, $P(A \text{ and } B)$. The probability of obtaining two heads in two flips of a coin can be expressed as $P(H_1 \text{ and } H_2)$, that is, obtaining a head on the first flip and a head on the second flip. The multiplication principle tells us that if there are two events and each event can occur in two ways, then there will be $2 \times 2 = 4$ possible outcomes: $H_1 T_2, H_1 H_2, T_1 H_2, T_1 T_2$. Also, the principle can be used to determine the number of ways of getting H_1 and H_2—1 (heads on first trial)$\times 1$ (heads on second trial). Hence $P(H_1 \text{ and } H_2) = \dfrac{1 \times 1}{2 \times 2} = \dfrac{1}{4}$. That is, there is a probability

of 1 out of 4 that when two coins are flipped, they will both land heads. We can write that finding as

$$P(H_1 \text{ and } H_2) = P(H_1) \cdot P(H_2) = \tfrac{1}{2} \times \tfrac{1}{2} = \tfrac{1}{4}$$

When two events, A and B, are *independent*—that is, when the number of ways of obtaining B is not affected by A, and vice versa—then, $P(A$ and $B) = P(A) \cdot P(B)$. When two outcomes are independent, the chance of both outcomes occurring is the product of their individual probabilities.

Suppose examination results of a test in school showed that 7 per cent of the students failed the English examination, 10 per cent failed the history examination, and 5 per cent failed both examinations. Are the two events "student fails English exam" and "student fails history exam" independent events? If they are, then we must have that P(failing both exams) = P(failing English exam)$\times P$(failing history exam). Our statistics stated that P(failing both exams) = .05, P(failing English exam) = .07, and P(failing history exam) = .10. Since $.05 \neq (.07)(.10)$, the events are not independent. Consider, if Bert buys a paper at Lil &Hy's luncheonette five days a week and Joan buys a paper four days a week at the same time and place, and the events are assumed independent, the chance of both buying the paper on the same day is $\tfrac{5}{7} \cdot \tfrac{4}{7} = \tfrac{20}{49}$, or 41 per cent. If it is found that they buy the paper at the same time and place around 80 per cent of the time, then the assumption of independence may not be true. That is, the events may be dependent upon each other.

We used the independence principle earlier when we determined the outcomes of flipping three coins (Figure 4.3.2), inasmuch as the outcome of flipping one coin is independent of flipping another coin. For example, $P(H_1 \text{ and } H_2 \text{ and } H_3) = \tfrac{1}{2} \times \tfrac{1}{2} \times \tfrac{1}{2} = \tfrac{1}{8}$. Do you see why "and" is used and not "or"? If we wanted to find the chance of obtaining two heads and one tail in three flips of a coin, we can satisfy this event in three ways: $H_1 H_2 T_1$, $H_1 T_1 H_2$, or $T_1 H_1 H_2$. Inasmuch as *any one* of these three arrangements satisfies the event, we have that

$$P(2 \text{ heads and 1 tail}) = P(H_1 H_2 T_1) + P(H_1 T_1 H_2) + P(T_1 H_1 H_2)$$

Now, to find $P(H_1 H_2 T_1)$ requires that we find $P(H_1)$ *and* $P(H_2)$ *and* $P(T_1)$—we use "and" since all three outcomes must occur in order to have

$H_1H_2T_1$. We then have that

$$P(2 \text{ heads and 1 tail}) = \tfrac{1}{2} \times \tfrac{1}{2} \times \tfrac{1}{2} + \tfrac{1}{2} \times \tfrac{1}{2} \times \tfrac{1}{2} + \tfrac{1}{2} \times \tfrac{1}{2} \times \tfrac{1}{2} = \tfrac{3}{8}$$

since $P(H_1H_2T_1) = \tfrac{1}{8}$, $P(H_1T_1H_2) = \tfrac{1}{8}$, and $P(T_1H_1H_2) = \tfrac{1}{8}$. Suppose that we had flipped a fair coin four times. What is the chance of obtaining two heads? Since we are flipping four coins, we must determine all the arrangements of two heads and two tails. If you work this out you will discover a very unexpected probability. Try it.

Many people use the multiplication principle intuitively, sometimes believing incorrectly that some outcome is due to occur because it has not happened for a long time. For example, when it has not rained for a while, people say "We're due for rain," or if a baseball player has not gotten a hit in a long time, people say, "He's due for a hit." It might soon rain and the player might soon get a hit, but it does not mean that these "due" events are due now. Over the long run, they will happen given that the conditions are similar to the conditions in the past that led us to have these beliefs.

We shall consider next an example that illustrates the common misunderstanding of independent versus dependent events. Suppose that we flipped a coin 10 times. The number of possible outcomes would be

$$\underbrace{2 \times 2 \times \cdots \times 2}_{10} = 1{,}024$$

If heads came up all 10 times,

$$P(H_1 \text{ and } H_2 \text{ and } \cdots \text{ and } H_{10})$$

$$= P(H_1) \cdot \cdots \cdot P(H_{10}) = \frac{1 \times 1 \times 1 \cdots \times 1}{1{,}024} = \frac{1}{1{,}024} = .00097$$

(this event should occur over the long run in 97 of 100,000 times). Again, we might believe that since this probability is so small, "tails is due" on the eleventh toss. However, on the eleventh toss, the probability of heads, $P(H) = \tfrac{1}{2}$. The unlikely event of 10 heads in a row has occurred; the eleventh toss is independent of the first 10 tosses. Indeed, over the long run, if the coin is fair, we should expect about an equal proportion of both heads and tails. But there is no reason to believe that it is going to begin evening out right now. Since a coin has no memory, it does not balance out outcomes—chance does that.

In addition to its use for independent events, the multiplication principle can also be used to determine the number of outcomes of events that are not independent, that is, events that are affected and "depend" on each other, called *dependent* events. For example, to determine the probability that three people have different birth dates involves dependent events, as each is dependent upon the preceding date. For, while the first person can have any of the 365 days, the second person can only have any of the 364 days, since one chance has been eliminated, the the third person can only have any of the remaining 363 days. Hence the probability that three people have different birth dates is

$$\frac{365 \times 364 \times 363}{365 \times 365 \times 365} \approx .99$$

Before reading the solution, jot down how many people you think would be needed to make the probability greater than 50 per cent that at least two people have the same birth date. We want to determine $P(B \geq 2)$, the chance that at least two people have the same birth date. We will find this probability when there are 23 people. Why 23? because when there are 23 people in the group, the probability of at least two having the same birth date $\approx .51$—more than 50 per cent!

The solution to this problem uses our second rule: $P(A) + P(A') = 1$. In place of "A" we will write "$B = 2$." We then have $P(B \geq 2) + P((B \leq 2)') = 1$. But $P((B \geq 2)')$ means $P(B < 2)$; that is, "not at least two" means "less than two," and "less than two" in this case "equals one," or each is a unique birth date. Thus $P(B \geq 2) + P(B = 1) = 1$. To solve for $P(B \geq 2)$, we subtract "$P(B = 1)$" from both sides, and arrive at $P(B \geq 2) = 1 - P(B = 1)$. From the earlier discussion we see that

$$P(B = 1; \text{ for 23 people}) = \frac{365 \times 364 \times 363 \times \cdots \times 343}{265 \times 365 \times 365 \times \cdots \times 365} \approx .49$$

Then $P(B \geq 2) = 1 - P(B = 1) = 1 - .49 = .51$.

Table 4.4.1 show some more surprising probability measures for other groups. Notice that as soon as there are 60 people gathered, there is a 99 per cent chance that at least two of them have the same birth date. Was your guess close? What is even more remarkable is that although 60 people produce a chance of 99 per cent, it takes 366 people to make it certain. Why?

Table 4.4.1. Birthday Problem

Group size	5	10	20	23	30	40	50	60
$P(B \geq 2)$	.03	.12	.41	.51	.71	.89	.97	.99

There is an interesting anecdote about how an incorrect understanding of independent events sent two people to jail. A married couple were convicted of a crime on the circumstantial evidence that the woman was white, blonde, and wore a ponytail while her husband was black, bearded, and drove a yellow car. The prosecution multiplied their estimates of the individual probabilities and arrived at the chance that 1 couple out of 12 million share this set of characteristics, and concluded that this couple, who was bound to be in the area of the crime, "most probably" was guilty. Later there was a reversal of the decision based on several factors: the probability estimates were never really measured in the real world—the occurrence of being a black male and being part of an interracial couple is not necessarily independent. Using the same method as the birthday problem, and the 12 million figure set by the prosecution, the judge was shown that there exists at least a 41 per cent chance that there is another couple in that area with the same characteristics as the convicted couple.

PROBLEMS

4.4.1 If you had four activities you wanted to do on a day off, how many different ways could you arrange to do them all? That is, how many different arrangements of the four activities, call them *A*, *B*, *C*, and *D*, would there be?

4.4.2 How many arrangements of lights are possible in the "Response to Stimulus Compound" experiment if only 0, 2, 4, or 8 lights could be on in any row and all three rows could not be zero?

4.4.3 A retailer receives two shipments of goods. It has been found in the past that 1.5 per cent (.015) of all shipments are spoiled in transit. Find the probability that:
   (a) Both shipments are spoiled.
   (b) Neither is spoiled.
   (c) Only one is spoiled (first is spoiled *and* not the second, or first is not spoiled *and* the second is).

4.4.4 If you had your keys in one of your four pockets, what is the chance that if you just picked a pocket at random, you would pick the correct pocket on the first pick; second pick; third pick; fourth pick?

4.4.5 In Texas, in 1956, 77 per cent of elementary school teachers were female, and 45 per cent of secondary school teachers were male.
   (a) What is the chance that a committee made up of one elementary school and one secondary school teacher contains no females?
   (b) What is the chance that the committee consisted of two females, or a female from the elementary school and a male from the high school?

4.5 Pascal's Triangle

If we wish to find the probability of obtaining two heads in five tosses of a coin—that is, the probability of two heads and three tails—we need to know how many arrangements of two heads and three tails are possible.

We can begin by listing them (e.g., *HHTTT, HTTHT*, etc.). This procedure takes considerable time and in the process we might easily omit some of the cases. Hence, we seek a more efficient way to determine the number of different outcomes. From the multiplication principle, we know that there are 32 possible quintuples of outcomes. What we do not know is the number of each of the different outcomes.

Blaise Pascal, French philosopher and mathematician (1623–1662), developed a simple method for determining the number of arrangements of each distinct outcome, where the outcomes consisted of two possibilities (e.g., heads or tails, defective or nondefective, toxic or nontoxic). This method, named after Pascal and for the triangular form it takes, is known as *Pascal's triangle* (Figure 4.5.1).

$n = 1$:				1				1		
$n = 2$:			1			2				1	
$n = 3$:		1		3				3		1	
$n = 4$:	1		4		6			4			1
$n = 5$:	1	5		10			10		5		1

Figure 4.5.1 Pascal's Triangle

The diagram has the following interpretation. If we had one trial ($n = 1$) and two possible outcomes (e.g., heads or tails), the number of ways of obtaining heads = 1 and the number of ways of obtaining tails = 1 (see row 1). If we had two trials, there would be one way of obtaining two heads (*HH*), two ways of obtaining a head and a tail (*HT, TH*), and one way of obtaining two tails (*TT*). The second row of the triangle shows the number of ways the different outcomes can occur: 1 2 1. To use the triangle, we need to know that in every row the first number (1) is the number of ways the outcomes would all be the same (e.g., in row $n = 3$, there is one way of getting all three coins to be heads). The second number in the row is the number of ways that one less outcome can occur (e.g., in row $n = 3$, $3 - 1 = 2$ heads can occur three times). Thus, the $n = 3$ row tells us that when we flip three coins, there will be one outcome of three heads, three outcomes of two heads and one tail, three outcomes of one head and two tails, and one outcome of all tails.

By studying Pascal's triangle, can you discover what the $n = 6$ row would look like? If we examine the $n = 5$ row and the row directly above it, $n = 4$, we can find some relationships. To begin with, the values at the ends of all the rows are 1. Second, "5" is below and between "1" and "4" in $n = 4$; and "10" is below and between "4" and "6." Check the rest of the triangle to see if, besides the 1's, every number is the sum of the two numbers that are diagonally above it. As this is the case, the $n = 6$ row should be

$$1 \quad (1+5) \quad (5+10) \quad (10+10) \quad (10+5) \quad (5+1) \quad 1$$

that is,

$$1 \quad 6 \quad 15 \quad 20 \quad 15 \quad 6 \quad 1$$

Hence, if we flipped a coin six times, there would be 1 arrangement of all six heads, 6 arrangements of five heads and one tail, 15 arrangements of four heads and two tails, and so on.

If we had to create every row in order to determine the next row, the procedure would be extremely laborious. There is a handy formula to find any of the number of arrangements for a two-outcome (binomial) situation:

$$\binom{n}{k} = \frac{n!}{k!(n-k)!}$$

The expression on the left side of the equal sign is written in a form to express that n is the number of outcomes possible (the number of the row) and k is the number of successes. For example, $\binom{6}{4}$ is the number of ways in $n = 6$ trials of containing four successes and two failures.

The fraction on the right-hand side contains the symbol "!" (read: factorial). Mathematicians use this symbol after a number to represent the product of all the natural numbers that lie between it and 1, inclusive. For example, $5! = 1 \times 2 \times 3 \times 4 \times 5$. The exclamation mark was chosen because of the magnitude of the product. For example,

$$25! = 15,511,000,000,000,000,000,000,000$$

Hence

$$\binom{6}{4} = \frac{6!}{4!(6-4)!} = \frac{6!}{4!\,2!} = \frac{1 \times 2 \times 3 \times 4 \times 5 \times 6}{1 \times 2 \times 3 \times 4 \times 1 \times 2} = 15$$

That is, in the $n = 6$ row, there are 15 arrangements containing four heads (successes) and two tails (failures). Similarly,

$$\binom{6}{5} = \frac{6!}{5!\,1!} = \frac{1 \times 2 \times 3 \times 4 \times 5 \times 6}{1 \times 2 \times 3 \times 4 \times 5 \times 1} = 6$$

that is, there are 6 arrangements of obtaining five heads and one tail in six tosses of a coin (*HHHHHT, HHHHTH, HHHTHH, HHTHHH, HTHHHH, THHHHH*). The $n = 6$ row can be written

$$\binom{6}{6} \quad \binom{6}{5} \quad \binom{6}{4} \quad \binom{6}{3} \quad \binom{6}{2} \quad \binom{6}{1} \quad \binom{6}{0}$$

in place of

$$1 \quad 6 \quad 15 \quad 20 \quad 15 \quad 6 \quad 1$$

Notice that $\binom{6}{6} = \binom{6}{0} = 1$. This equation states that in six trials there is one way of obtaining six successes or zero successes (six failures): *SSSSSS* or *FFFFFF*. In other words, the number of ways of obtaining six successes or six failures in six trials is the same: one. Notice that this is not the probability of this occurrence but the number of possible arrangements. From the formula, $\binom{6}{6} = \dfrac{6!}{6!\,0!} = 1$; thus, we agree to define $0! = 1$ so that the formula holds in this case also.

Now we can find the number of outcomes for any particular case of a binomial distribution. For example, if we had sent out $n = 27$ questionnaires and we wanted to know how many ways we could obtain 25 positive replies and 2 negative replies, we do not have to create the $n = 27$ row. The answer would be $\binom{27}{25} = \dfrac{27!}{25!\,2!} = 351 = \binom{27}{2}$. That is, $\binom{27}{25} = \binom{27}{2}$, since the number of ways of obtaining 25 positive responses out of 27 responses is equal to the number of ways of obtaining 2 negative responses out of 27 responses. For when there are 2 negative responses, there must be 25 positive ones if there are 27 responses.

P R O B L E M S

4.5.1 (a) Show that $\binom{6}{3} = 20$.

(b) Create the $n = 7$ row.

(c) Find the number of total outcomes for each of the first five rows. Create a formula that relates the number of trials, n, to the total number of outcomes. Verify your formula using the multiplication principle and check it for the $n = 6$ and 7 rows.

(d) Using the definition of $\binom{n}{k}$, show that $\binom{7}{4} = \binom{6}{3} + \binom{6}{4}$.

4.5.2 If a drug is to be given to 4 of 10 people, how many different groups of 4 people could get the drug? How many different groups of 6 people would not get the drug?

4.6 Binomial Distribution

The coefficients (constants) found in Pascal's triangle tell us the number of ways in which we get a particular set of outcomes when a binomial experiment is performed a certain number of times. For example, in $n = 3$ trials, the coefficients of outcomes will be 1   3   3   1, no matter what the binomial experiment. However, if the individual probabilities differ for two different binomial experiments while the number of arrangements possible are the same, the outcome probabilities will be different. For example, if $P(\text{heads}) = P(\text{tails}) = .50$, while $P(\text{relief}) = .90$ and $P(\text{nonrelief}) = .10$, we would not expect the $P(2 \text{ heads and } 1 \text{ tail}) = P(2 \text{ relief and } 1 \text{ nonrelief})$, although each of these events can occur in three ways. The coefficients in the triangle tell us in how many ways an event can occur—not the probability of the event.

In $n = 3$ trials, we can obtain two heads and one tail in three ways: *HHT*, *THH*, *HTH*; $P(HHT) = P(H) \cdot P(H) \cdot P(T) = \frac{1}{2} \cdot \frac{1}{2} \cdot \frac{1}{2} = \frac{1}{8}$. If we rearranged the two heads and one tail, the probability of each of the three outcomes would still be $\frac{1}{8}$. Hence $P(2 \text{ heads and } 1 \text{ tail in any order}) = 3 \cdot \frac{1}{8} = \frac{3}{8}$. The coefficient "3" is the number of ways of arranging the outcomes. We can write "$P(HHT)$" as "$(P(H))^2 \cdot P(T)$." Thus $P(2 \text{ heads and } 1 \text{ tail in } any \text{ order}) = 3(\frac{1}{2})^2(\frac{1}{2}) = \frac{3}{8} = .375$. While in $n = 3$ trials we can also obtain two relief and one nonrelief in three ways: *RRR'*, *R'RR*, *RR'R*; however, $P(RRR') = P(R) \cdot P(R) \cdot P(R') = \frac{9}{10} \cdot \frac{9}{10} \cdot \frac{1}{10}$. Hence $P(2 \text{ relief and } 1 \text{ non-relief } in \text{ any order}) = 3(.9)^2(.1) = \frac{243}{1,000} = .243$. So, if we flipped three coins 1,000 times, we would expect around $.375(1,000) = 375$ times, assuming a fair coin, that 2 heads and 1 tail would occur. Whereas, if we took 1,000 samples of three people, assuming that $P(\text{relief}) = .90$, we would expect about $.243(1,000) = 243$ cases in which where two of the three people would have positive reactions. Table 4.6.1 lists the probabilities for each of these possible outcomes.

Table 4.6.1. Binomial Outcomes and Associated Probabilities for $n = 3$ Trials

Number of heads	P(heads)	Number of relief	P(relief)
3	$\binom{3}{3}(.50)^3(.50)^0 = .125$	3	$\binom{3}{3}(.90)^3(.10)^0 = .729$
2	$\binom{3}{2}(.50)^2(.50)^1 = .375$	2	$\binom{3}{2}(.90)^2(.10)^1 = .243$
1	$\binom{3}{1}(.50)^1(.50)^2 = .375$	1	$\binom{3}{1}(.90)^1(.10)^3 = .027$
0	$\binom{3}{0}(.50)^0(.50)^3 = .125$	0	$\binom{3}{0}(.90)^0(.10)^3 = .001$
	$\overline{1.000}$		$\overline{1.000}$

Note that the shapes of the distributions are dependent upon the probability of success. However, the sum of the individual probabilities for each binomial distribution is 1, as we have included all the possibilities.

Each of the terms for any n can be found by creating the terms of the binomial expansion. For example, if $n = 3$,

$$(A + B)^3 = (A + B)(A + B)(A + B) = 1 \cdot A^3 + 3A^2B + 3A^1B^2 + 1 \cdot B^3$$

The coefficients are the number of ways of arranging each term, and if we replace "A" by "$P(A)$" and "B" by "$P(B)$," we have the binomial probability distribution when $n = 3$:

$$1 \cdot (P(A))^3 + 3 \cdot (P(A))^2 \cdot P(B) + 3 \cdot (P(A))^1 (P(B))^2 + 1(P(B))^3$$

Hence, if $P(A) = .50$, or $P(A) = .90$, we have all the terms that we found in Table 4.6.1.

The general formula for the binomial probability value for k successes in n trials, where the probability of success is p, and the probability of failure is $1 - p$, is

$$\binom{n}{k} p^k (1-p)^{n-k} = \frac{n!}{k!\,(n-k)!} p^k \cdot (1-p)^{n-k}$$

For example, if $n = 4$, $p = .20$, and $1 - p = .80$, we have the probability distribution of Table 4.6.2.

Table 4.6.2. Binomial Probability Distribution for $n = 4$ and $p = .20$

k	Number of arrangements, $\binom{n=4}{k}$	P(one arrangement) $p^k(1-p)^{n-k} = (.20)^k(.80)^{n-k}$	P(all arrangements) $\dfrac{n!}{k!\,(n-k)!}p^k(1-p)^{n-k}$
4	1	$(.20)^4(.80)^0 = .0016$	$1(.20)^4(.80)^0 \doteq .0016$
3	4	$(.20)^3(.80)^1 = .0064$	$4(.20)^3(.80)^1 = .0256$
2	6	$(.20)^2(.80)^2 = .0256$	$6(.20)^2(.80)^2 = .1536$
1	4	$(.20)^1(.80)^3 = .1024$	$4(.20)^1(.80)^3 = .4096$
0	1	$(.20)^0(.80)^4 = .4096$	$1(.20)^0(.80)^4 = .4096$
		$.5456$	1.0000

PROBLEMS

4.6.1 A married couple has four children. Assuming that P(boy) $= P$(girl), find the probability that:
  (a) Only one child is a boy (and the other three are girls).
  (b) Two are boys.
  (c) Three are girls.
  (d) At least three are girls.
  (e) All are boys.

4.6.2 Suppose a fair coin is flipped four times. Find the probability that:
  (a) All coins fall tails.
  (b) Two are heads and two are tails.
  (c) At least one is a head.

4.6.3 Create a relative frequency histogram of the following:
  (a) Table 4.6.1—you might want to switch the order of the number of successes from 4–0 to 0–4 on the horixontal axis of the histogram.
  (b) Table 4.6.2.
  (c) The binomial probability distribution, where $n = 5$ and $p = .50$.
  (d) What are the shapes of the distributions of the "relief" section of Table 4.6.1 and the one that you created in part (c)?

4.6.4 What is the probability of hitting a bulls-eye on a dartboard three time in a row if the probability of hitting the bulls-eye once is .10? Hitting it twice in three tries? At least once? At most once? (NOTE. First create a tree diagram—assume independence.)

4.6.5 Is it always true that the more chances one gets, the better the chance of one's success? (Why?)

4.7 Mathematical Expectation

In Section 4.6 we found the probabilities of outcomes in a binomial probability distribution. One very valuable concept examined earlier, the mean, can also be incorporated into a discussion of probability: that given a certain set of outcomes and their associated probabilities, the question arises as to what we might expect to have occur over the long run. For example, if a set of grades of a student in a statistics class was 75, 80, 75, 85, 90, 75, 70, 80, 85, and 90, we could represent these 10 scores and their number of occurrences as listed in Table 4.7.1. We see that 70 occurred $\frac{1}{10}$ of the time, 75 occurred $\frac{3}{10}$ of the time, and so on. We can find the mean of the set of scores $\bar{s}$, as we would find the arithmetic mean:

$$\bar{s} = 70(\tfrac{1}{10}) + 75(\tfrac{3}{10}) + 80(\tfrac{2}{10}) + 85(\tfrac{2}{10}) + 90(\tfrac{2}{10})$$
$$= 80.5$$

We would say that we expect this student to be an "80 student" over the long run.

Table 4.7.1. Relative Frequency Distribution of Grades

S	70	75	80	85	90
$P(S)$	$\frac{1}{10}$	$\frac{3}{10}$	$\frac{2}{10}$	$\frac{2}{10}$	$\frac{2}{10}$

Another example might be helpful in considering the mean of relative frequency and probability distributions. Consider Figure 4.7.1, which

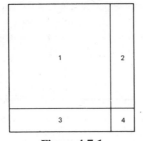

Figure 4.7.1

depicts a new type of dartboard. The "1" region contains 50 per cent of the total area, the "2" region 20 per cent, the "3" region 20 per cent, and the "4" region 10 per cent.

If the game were played by tossing a dart at the board without looking, we would expect that 50 per cent of the time 1 point will be scored (hit in the "1" region), 20 per cent of the time 2 points will be scored, and so on. These probability values are determined by the logic of the situation, not by relative frequency—although the relative frequencies, over the long run, should approach these probability values. We could represent these possible scores and their corresponding probabilities as shown in Table 4.7.2.

Table 4.7.2. Dart Scores and Associated Probabilities

S	1	2	3	4
$P(S)$	.50	.20	.20	.10

As we would determine the mean, we can determine the *mathematical expectation* of the scores. That is, we use the mean to determine the expected value when we have determined the relative frequency values. When we determine the probability values, we interpret the mean as the mathematical expectation. The mean value moves toward the mathematical expectation as the sample size increases. In this situation, the mathematical expectation, $E(S) = 1(.50) + 2(.20) + 3(.20) + 4(.10) = 1.90$. We can say that the average score per dart throw is 1.90. Suppose that we took 1 per cent of the "2" region (leaving 19 per cent) and created a "5" region containing only 1 per cent of the total area. But instead of scoring 5 points for this region, assign it a value of 100. Then we would expect, over the long run, that we would obtain a score of 2.88 per throw, as $E(S) = 1(.50) + 2(.19) + 3(.20) = 4(.10) + 100(.01) = 2.88$. Notice that although we will hit the 100-point region 1 per cent of the time over the long run, the fact that it is possible has raised our mathematical expectation over 50 per cent. However, we shall hit the "jackpot" only 1 per cent of the time on the average. Amusement-park games set up situations similar to this one. If the original figure was redrawn so that each region contained 25 per cent of the area, what would the mathematical expectation be? Try it. Should the $E(S)$ be larger or smaller than 1.90? Why? Does it make sense to you that this value should be greater than 1.90?

We can now define mathematical expectation, $E(X)$. The mathematical expectation for any set of outcomes having values $x_1, x_2, \ldots, x_n$ with associated probabilities $P(x_1), P(x_2), \ldots, P(x_n)$ is

$$E(X) = x_1 P(x_1) + x_2 P(x_2) + \cdots + P(x_n)$$

PROBLEMS

4.7.1 Find the expected number (mathematical expectation) of heads in Table 4.3.2.

4.7.2 Find the expected number of girls in a family with six children.

4.7.3 Find the expected number of heads and the number of relief from Table 4.6.1.

Suppose that we had a three-question match-up test. We have to match the following people: John F. Kennedy, Martin Luther King, Jr., and Doris Lessing to their correct years of birth: 1917, 1929, and 1919. If we guessed at the answers, what would be the expected number of correct guesses? Inasmuch as the number of correct responses are 0, 1, 2, and 3, we need to determine the associated probabilities in order to find the expected number of correct guesses.

The chance of assigning Kennedy to the right year is 1 out of 3. Once we have done that, there are two people and two dates left. Thus the chance of assigning King to his birth year is 1 out of 2. And finally, Lessing must be assigned to the only remaining date with the probability of 1 out of 1. Hence the probability of getting all three birth years correct is $\frac{1}{3} \times \frac{1}{2} \times \frac{1}{1} = \frac{1}{6}$; that is, $P(\text{correct} = 3) = \frac{1}{6}$. To find $P(C = 2)$, that is, that two people get assigned to the right year, suppose that King and Lessing are assigned correctly. Then in order to get two correct, we must get one incorrect. But Kennedy can only be assigned to the one remaining year, and that year must be the correct one, as it is the only one left. Thus $P(C = 2) = 0$. It cannot happen, for, when we guess two correct, we must necessarily guess all three correct. In order to get only one correct, two must be assigned to the incorrect dates. Suppose that we guess King's birth year correctly. By pure chance this will occur $\frac{1}{3}$ of the time. From the two remaining years, we want to assign Lessing to the incorrect year; this occurs, by chance, $\frac{1}{2}$ of the time. And Kennedy must be assigned to the remaining year incorrectly, with a probability of $\frac{1}{1}$. Hence the probability of picking only King correct is $\frac{1}{3} \times \frac{1}{2} \times \frac{1}{1} = \frac{1}{6}$. Since we want the chance of picking any one of the three correctly, we have $P(C = 1) = P$ (King correct or Kennedy correct or Lessing correct) $= \frac{1}{6} + \frac{1}{6} + \frac{1}{6} = 3 \times \frac{1}{6} = \frac{3}{6}$. To find. $P(C = 0)$, note that $P(C = 0) + P(C = 1) + P(C = 2) + P(C = 3) = 1$. Thus $P(C = 0) + \frac{3}{6} + 0 + \frac{1}{6} = 1$ implies that $P(C = 0) = \frac{2}{6}$. It is interesting to note that in this situation it is harder to guess none correct than one correct; that is, $P(C = 0) = \frac{2}{6}$ is less than $P(C = 1) = \frac{3}{6}$. The number correct and their associated probabilities appear in Table 4.7.3. When we guess, we should expect to obtain $E(C) = 0 \times \frac{2}{6} + 1 \times \frac{3}{6} + 3 \times \frac{1}{6} = 1$ correct.

Table 4.7.3. Three-Question Match

C	0	1	3
$P(C)$	$\frac{2}{6}$	$\frac{3}{6}$	$\frac{1}{6}$

Games of chance run for a profit are not fair. A game is fair if and only if the net expected gain is zero; that is, the player is expected to break even. Although the New York State Lottery does not offer the expectation for everyone to break even, the money does go toward education. The following payoff in this lottery for a 50 cent ticket is

1. All six digits correct in order wins $50,000.
2. First or last five in order wins $2,000.
3. First or last four in order wins $125.
4. First or last three in order wins $25.
5. First or last two in order wins $2.50.

This prize schedule is for each 1 million tickets sold. The possible outcomes are $50,000, $2,000, $125, $25, $2.50, and −$.50 (the cost of the ticket). We need to determine the probabilities associated with each outcome.

The multiplication principle can be used to find the total number of arrangements of a number that is six digits in length. Since any of the 10 digits 0 to 9 can fill any of the six digits, there are $10 \times 10 \times 10 \times 10 \times 10 \times 10 = 1,000,000$ six-digit possible numbers. Of these, only one will win the big prize. Hence $P(W = 50,000) = \frac{1}{1,000,000}$. To obtain the first five digits correctly means that the sixth digit must be incorrect. So there are $1 \times 1 \times 1 \times 1 \times 1 \times 9$ ways of getting the first five correct, and similarly for the last five $(9 \times 1 \times 1 \times 1 \times 1 \times 1)$. Hence $P(W = 2,000) = \frac{18}{1,000,000}$. To obtain only the first four correct means that the first four are correct, the fifth must be incorrect, and the sixth may be correct. So there are $1 \times 1 \times 1 \times 1 \times 9 \times 10 = 90$ ways of getting the first four in a row correct. Note that the "10" in the sixth place implies that all the digits can fill in that spot. Since the fifth digit is wrong, it makes no difference what number falls in the last spot. And obtaining the last four correct has the same product. Hence $P(W = 125) =$

$\frac{180}{1,000,000}$. You should try to show that $P(W=25)=\frac{1,800}{1,000,000}$ and $P(W=2.50)=\frac{40,419}{1,000,000}$ (this is really a tricky one). The chance of losing the wager, $P(W=-.50)$ is found by adding up the number of winning tickets and subtracting from 1,000,000. Hence $P(W=-.50)=\frac{957,582}{1,000,000}$.

The expected net gain can be determined from Table 4.7.4.

Table 4.7.4. Expected Net Gain for N.Y. State Lottery

W	50,000	2,000	125	25	2.50	−.50
$P(W)$	.000001	.000018	.00018	.0018	.040419	.957582

The expected net gain is

$$E(W) = 50,000(.000001) + 2,000(.000018) + 125(.00018)$$

$$+ 25(.0018) + 2.50(.040419) - .50(.957582)$$

$$= .05 + .036 + .0225 + .0450 + .1010475 - .478791 = -.22$$

So, for every 50 cents bet, 22 cents is taken. Fortunately, the money is going for a good cause.

P R O B L E M S

4.7.4 Compare the expected number of heads in Tables 4.3.1 and 4.3.2.

4.7.5 What is the expected number of girls born in four births from Problem 4.3.1?

4.7.6 What is the expected number of times of hitting the bulls-eye in Problem 4.6.4?

4.7.7 If S is the sum of the face numbers when two dice are thrown, and $P(S)$ is the probability of obtaining this sum such that:

S	2	3	4	5	6	7	8	9	10	11	12
$P(S)$	$\frac{1}{36}$	$\frac{2}{36}$	$\frac{3}{36}$	$\frac{4}{36}$	$\frac{5}{36}$	$\frac{6}{36}$	$\frac{5}{36}$	$\frac{4}{36}$	$\frac{3}{36}$	$\frac{2}{36}$	$\frac{1}{36}$

(a) Find the expected sum.
(b) Make a bar graph with S on the horizontal axis and $P(S)$ on the vertical axis. What type of distribution does this bar graph appear to have?
(c) Find $P(S$ is even$)$; $P(S \geq 10)$; $P(3 \leq S \leq 6)$; $P(3.1 \leq S \leq 7.8)$.

4.7.8 How much should a lottery ticket cost if the game would be fair; $E(W)=0$?

4.8 Summary

In this chapter we explored the concept of probability and how we can use it to evaluate situations and make decisions. Chance based on experience is referred to as relative frequency. As one collects more and more data, the relative frequency "zeros-in" to a fractional value called the probability of the event. When we are able to theoretically determine the chance of an event happening, we do not have to depend upon examining sample data. However, when the data do not approximate what we expect to occur, we have to reexamine the origin and method of collecting the data and the assumptions that led us to have certain expectations. If we understand a probabilistic situation, we can predict what will happen. The third model we created in the "response" problem determined a collection of probability values that closely approximated the corresponding collection of relative frequency values of responses. Hence, for the nine test patterns considered, we can say that model 3 demonstrates an understanding of how individuals react. However, since there are 63 possible test patterns (see Problem 4.4.1), we cannot guarantee that model 3 will continue to be a good approximation for the 54 test patterns that were not considered.

The multiplication principle, along with the probability rules we created, aided us in determining the number of outcomes and the probability of compound events ("*A* or *B*" and "*A* and *B*"). With this knowledge, we examined the binomial probability distribution, a theoretical distribution that plays an important role in probability and statistics, as we will see. When we can assign a set of associated probabilities to a set of outcomes of an experiment, we are able to determine the mathematical expectations of what we should expect to occur over the long run. We will use these concepts in our examination of inferential statistics.

4.9 Review of Formulas

I. Probability of an event A:

$$P(A) = \frac{\text{number of simple outcomes favorable to } A}{\text{total number possible simple outcomes}}$$

II. Properties and rules for probability
A. $0 \le P(A) \le 1$, for any event A.
B. $P(A \text{ or } A') = 1$, for any event A.
C. $P(A \text{ or } B) = P(A) + P(B)$, when A and B have no outcomes in common, for any events A and B.
D. $P(A \text{ or } B) = P(A) + P(B) - P(\text{any outcomes common to both } A \text{ and } B)$.
E. $P(A \text{ and } B) = P(A) \cdot P(B)$, if and only if events A and B are independent.

III. In n trials the number of ways of obtaining k successes and $n-k$ failures is

$$\binom{n}{k} = \frac{n!}{k!\,(n-k)!} \qquad \text{where } k! = 1 \times 2 \times \cdots \times k$$

IV. The chance of obtaining k successes in n trials, where the probability of success on each trial is p, there are $n-k$ failures, and the probability of failure on each trial is $1-p$, is

$$\binom{n}{k} p^k (1-p)^{n-k} = \frac{n!}{k!\,(n-k)!} p^k (1-p)^{n-k}$$

V. If a collection of outcomes $x_1, x_2, \ldots, x_n$ have associated probabilities $P(x_1), P(x_2), \ldots, P(x_n)$, the mathematical expectation for the outcomes is

$$E(x) = x_1 \cdot P(x_1) + x_2 \cdot P(x_2) + \cdots + x_n \cdot P(x_n)$$

4.10 Problem Set

4.10.1   Automobiles can be classified according to size as compact, standard, and luxury. What would you guess is the relative frequency distribution of sales for those classifications? Contact an automobile dealership in your area and compare your findings with theirs. Compare your findings with those of your class—can you make any generalizations?

4.10.2   From your experience (consider Problem 4.2.1), predict the relative frequency of a tack landing on its side, and compare it to the one found by recording the outcomes from tossing 50 tacks (or 10 tacks five times, etc.).

4.10.3   *Class Project* (needs electronic buffs): Create the light board discussed in the "Response" problem. Choose six other test patterns and, using the three models, determine the probability of occurrences. Then find a random sample of 50 students and record the relative frequency of their responses to the six test patterns. How do they compare to the models' predictions?

4.10.4   If $P(T) = .20$, $P(E) = .40$, and $P(H) = .40$, find the probability of picking, in order, the letters that spell:
(a) He.
(b) Et.
(c) The.
(d) Tee.
(e) Repeat (a) to (d) if the order does not count.

4.10.5 The number of students, by sex and class, who are taking a mathematics course in the fall term, is shown in the following table:

		Sex
Class	Male	Female
Freshman	160	140
Sophomore	120	110
Junior	100	90
Senior	80	70

What is the probability that a student chosen at random from the group of students taking a mathematics course will be
(a) A female?
(b) A junior?
(c) A male sophomore?
(d) A male who is at most a sophomore?
(e) A female or a freshman?

4.10.6 On Monday night, there are two hourly television shows that Jibby and Rose like to watch. But, as this is the season for repeats, they will not watch a repeat. If they have seen 36 of the 40 shows of show *A* and 32 of the 40 shows of show *B*, what is the chance that they will watch:
(a) Both shows tonight?
(b) Only show *A*?
(c) Only one of them?
(d) Neither one?

4.10.7 Two good flashlight batteries get mixed in with three bad ones. You plan to test all the batteries until you find both good ones.
(a) Construct a tree diagram for this experiment.
(b) What is the probability that the second good battery you find is the second tested? Third tested? Fourth tested? Fifth tested?
(c) Sum the probabilities that you found in part (b). Does the sum make sense? Why?
(d) How many arrangements of the five batteries are possible?

4.10.8 Find:
(a) How many ways can three successes and five failures occur in eight trials?
(b) How many ways can two successes and six failures occur in eight trials?

(c) Use parts (a) and (b) to determine how many ways three successes and six failures can occur in nine trials.

(d) (For algebra buffs.) We saw that in Pascal's triangle any coefficient could be determined by summing the two coefficients that lie diagonally above the coefficient we are looking for. In general, this can be written

$$\binom{n+1}{k} = \binom{n}{k} + \binom{n}{k-1}$$

Prove this. But first try to prove it, using the definition of $\binom{n}{k}$ for the case where $n = 4$ and $k = 3$.

4.10.9 A basketball player has a free-throw percentage of .75. Keeping this same percentage for each throw, determine the chance that in four attempts he makes:
(a) 0 baskets.
(b) 1 basket.
(c) 2 baskets.
(d) 3 baskets.
(e) 4 baskets.
(f) What is the expected number of baskets that he will make in four attempts?

4.10.10 Using the data from Problem 4.10.9, find the chance that he makes:
(a) Between 2 and 4 baskets (inclusive).
(b) At most 2 baskets.
(c) Between 0 and 4 baskets (inclusive).
(d) At least 3 baskets.
(e) Can you make an equation out of your findings for parts (b), (c), and (d)?

4.10.11 Another member of the team has a free-throw percentage of .80. How many baskets should he expect to make in four attempts?

4.10.12 A roulette wheel in Nevada has 38 equally likely outcomes: 0, 00, 1, 2, . . . , 36. The numbers 0 and 00 are green, while the 36 other numbers are divided evenly into red and black. Determine the mathematical expectation when the possible outcomes are
(a) 0, 00, . . . , 36—where you bet on one number and if it comes up, you are paid 36 units, including the unit you bet.
(b) Green, red, black—where you bet on red or black and if it comes up you are paid 2 units, including the unit you bet.
(c) Do parts (a) and (b) for the European roulette system, where 00 is not on the board (so that there are only 37 possibilities).

Chapter 5

Applications of the Binomial Probability Distribution

5.1 Introduction

Whenever we rely on rules or a model to help us make decisions, we have to be sure that the rules or model apply. For example, some rules for preparing to run in a race might be: "Get plenty of rest, take vitamins, and exercise lightly." Would you include as a rule, "Eat a bowl of spaghetti an hour before the race"? Some runners do. And there are doctors who claim that the spaghetti, loaded with carbohydrates, will provide added energy in a sprint race! If you were the manager of a supermarket and had to decide how much space to allow for each brand of toothpaste, would you use the following decision model: "5 per cent of the space should be given to brands that make 5 per cent profit; 10 per cent of the space for 10 per cent profit makers; and so on"? What would be the rule for a brand that makes 100 per cent profit? Is it not possible that some brands, while they make less profit per sale, sell so well that their total profit is greater than the total profit of an infrequently bought, high-profit brand? Therefore, a profit and demand model seems to be a better "fit" than just a profit model. In this chapter we shall examine characteristics of the binomial probability distribution and look at situations in which the binomial probability model can be useful in making decisions.

5.2 Binomial Probability Distribution

In Chapter 4 we began our discussion of the binomial probability distribution and found that it was very useful in determining certain probabilities of outcomes. In this chapter we shall use the binomial probability distribution to aid us in making decisions. We can use this model if the following conditions are met:

1. There are only two possible types of outcomes.
2. The number of trials is known.
3. The probability of success on each trial is the same.
4. The trials are independent of each other.

If these four assumptions are satisfied, we can determine that the probability of obtaining x "successes" in n trials, where the probability of success is p and the probability of failure is $q = 1 - p$, is

$$P(x \text{ successes}) = \binom{n}{x} p^x q^{n-x} = \frac{n!}{x!\,(n-x)!} p^x q^{n-x}$$

For example, if Michael had watched a certain television series and enjoyed 70 per cent of the shows and disliked the remaining 30 per cent, the chance that he would like four of the next five shows can be found by letting $n = 5$, $x = 4$, $p = .70$, and $q = .30$, such that

$$P(x = 4) = \binom{5}{4}(.70)^4(.30)^1 = \frac{5!}{4!\,1!}(.70)^4(.30)^1 = 5(.240)(.30) = .360$$

That is, there is about 36 per cent chance that he would like four of the next five shows of the series. The binomial probability model can be used because the four requirements were all satisfied: there are only two types of outcomes—like or dislike; the number of shows is known to be 5; the probability of success is constant at .70; and each show is independent of the next show. Similarly, the chance of enjoying 0, 1, 2, 3, or 5 of the next five shows is found by

$$P(x = 0) = \frac{5!}{0!\,5!}(.70)^0(.30)^5 = .002$$

$$P(x = 1) = \frac{5!}{1!\,4!}(.70)^1(.30)^4 = .028$$

$$P(x = 2) = \frac{5!}{2!\,3!}(.70)^2(.30)^3 = .132$$

$$P(x = 3) = \frac{5!}{3!\,2!}(.70)^3(.30)^2 = .309$$

$$P(x = 5) = \frac{5!}{5!\,0!}(.70)^5(.30)^0 = .168$$

If the next 10 shows were "not enjoyed," consideration would have to be given to whether to keep or change the probability of enjoyment value of $p = .70$. It is possible that chance alone could be the cause of the 10 poor shows; in fact, he could find 70 per cent of the shows to his liking if he continued watching. It is also possible that the quality of the show had gone down and, as such, the chance of enjoying the show should be lowered from 70 per cent. A similar situation exists in machine production. Suppose that a machine made electronic components for television sets. If the machine was said to produce 2 per cent defective parts over the long run, how many samples containing 5 per cent defective parts would have to be found before deciding that the "overly" defective parts are not due to chance but rather to machine failure? We shall examine similar situations shortly, using the binomial probability distribution to aid our decision making.

Graphically, we can represent the binomial probability distribution for liking the five television shows as in Figure 5.2.1. Notice that the number of

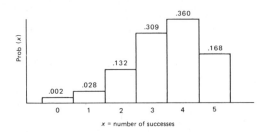

Figure 5.2.1 Binomial Probability Distribution for $n = 5$, $p = .70$

successes are located at the midpoints of the bases of the rectangles. Each rectangle represents the probability measure, and the sum of the areas of the rectangles is 1, since all the outcomes have been accounted for. (Actually, the sum in this situation is .999, since the probabilities have been rounded off to three places.) The chance that at least three of the five shows will be enjoyed is found by determining $P(x = 3) + P(x = 4) + P(x = 5) = .837$. Similarly, the chance that at most two shows will be enjoyed is $P(x = 0) + P(x = 1) + P(x = 2) = P(x \leq 2) = .162$.

How many of these five shows should Michael expect to enjoy? Recall: mathematical expectation was defined to be the sum of the products of the values of the outcomes and their associated probabilities:

$$E(X) = x_1 \cdot P(x_1) + x_2 \cdot P(x_2) + \cdots + x_n \cdot P(x_n)$$

In this situation, $x_1 = 0$ (none enjoyed), $x_2 = 1, \ldots, x_6 = 5$, and $P(x_1) = .002$, $P(x_2) = 0.28, \ldots, P(x_6) = .168$. We then obtain:

$$E(X) = 0(.002) + 1(.028) + 2(.132) + 3(.309) + 4(.360) + 5(.168)$$

$$= 3.5$$

Thus the expected number of enjoyable shows, where $n = 5$ and $p = .70$, is 3.5. It appears in this situation that not only does $E(X) = 3.5$, but so does $np = 5(.70) = 3.5$. Does this seem reasonable to you?—that the expected outcome is np, where p is the probability of success and n is the number of trials?

Rather than have to calculate probabilities for each binomial situation, we can use Table II in Appendix B. For example, to find the chance that in a random sample of $n = 20$ resistors, there are 3 defectives when the manufacturer stated that the probability of a defective $P(d) = .05$, look in

the n column to find $n = 20$ and $x = 3$, and across to the "$p = .05$" column
to find .060. This tells us that $\dfrac{20!}{3!\,17!}(.05)^3(.95)^{17} = .060$. That is, there is a
6 per cent chance of obtaining 3 defectives in a sample of 20 when
$P(d) = .05$—a small chance, but still possible.

P R O B L E M S

5.2.1 Reconsider Problems 4.7.2 to 4.7.4 and see whether $E(X) = np$ in
those situations. Are they all binomial situations?

5.2.2 Suppose that Susan was absent for the material to be covered on a
10-question true–false test. As she was adventurous, she decided to
take the test and guess at the answers; thus, P(guessing a correct
answer) $= .50$. Determine:
(a) If this situation fits a binomial probability model.
(b) P(3 correct).
(c) P(5 correct).
(d) P(10 correct).
(e) Expected number of correct. (Does it equal np?)
(f) The graph of the probability distribution using the number of
correct responses as class midpoints.

5.2.3 How could you classify the distribution in Figure 5.2.1?

You should have discovered that in Problems 5.2.1 and 5.2.2(e), the
expected number, the mean, is in fact equal to np, where n is the number of
trials and p the probability of success. All these situations fitted the
binomial model. Two formulas that can be proved, but which will not be
derived here, are the formulas for the mean and standard deviation of a
binomial probability distribution: $\mu = np$ and $\sigma = \sqrt{npq}$. In Problem 5.2.2,
since $p = .50$ and $n = 10$, we find that $\mu = (10)(.50) = 5$. This is reasonable
to expect; since Susan is guessing, and by pure chance she will guess as
often correctly as incorrectly, we would expect that over the long run she
would obtain 5 of the 10 correct. The standard deviation, $\sqrt{npq} =$
$\sqrt{(10)(.5)(.5)} = \sqrt{2.5} = 1.58$ (Table I in Appendix B). The mean is the
expected outcome, and the unit of variability we use to describe the
distribution of scores about the mean is the standard deviation. Thus within
1 standard deviation of the mean, $(\mu - 1\sigma, \mu + 1\sigma) = (3.42, 6.58)$, lie 65.6
per cent of the scores. This is due to the fact that scores of 4, 5, and 6 lie in
the 1σ interval, and $P(x = 4) + P(x = 5) + P(x = 6) = .656$. Over the long
run, a student who is guessing on a 10-question true–false test will obtain
either 4, 5, or 6 correct about 65.6 per cent of the time. If a score of at least

6 is passing, the chance the student will pass by guessing is

$$P(x \geq 6) = P(x = 6) + P(x = 7) + P(x = 8) + P(x = 9) + P(x = 10)$$
$$= .377 \quad \text{(Table II)}$$

So if 100 10–question quizzes were given, a student who guessed all the time would probably pass about 38 quizzes. We should emphasize "Probably pass about," because although it is possible that the student could pass none or even all the quizzes, the expected outcome is about 38.

As we have just seen, in order to determine what per cent of the scores fall within 1 or 2 standard deviations from the mean, we need to calculate the individual probabilities. When n is small, the computations are readily available from the table. But when n is large, the tables are not helpful. For example, if $n = 100$, $p = .30$, and $q = .70$, then $\mu = 30$ and $\sigma = \sqrt{100(.30)(.70)} = 4.58$. We can create the interval that is 1 standard deviation of the mean: $(30 - 4.58, 30 + 4.58) = (25.42, 34.58)$. But we do not know the per cent of the distribution within 1 standard deviation of the mean since we have not calculated $P(x = 26) + P(x = 27) + \cdots + P(x = 34)$. Fortunately, when n is large, we can approximate the binomial distribution with the normal distribution (bell-shaped curve). That is, as n increases, the shape of the binomial distribution approaches the shape of the normal distribution. And as we will see in Chapter 6, within 1 standard deviation of the mean of a normal distribution there is approximately 68 per cent of the area, within 2 standard deviations of the mean there is approximately 95 per cent of the area, and within 3 standard deviations of the mean there is approximately 99.7 per cent of the area. In the particular case we were just considering, our normal distribution approximation is that 68 per cent of the area is contained within the 1 standard deviation interval. From more extensive binomial probability tables we find the exact area to be .674. As you can see, the approximation is quite good. More will be said about this relationship later.

PROBLEMS

5.2.4 (a) Find the mean and standard deviation of the distribution in Problem 5.2.2.

(b) Determine the per cent of scores that are within 2σ of the mean.

(c) Is the normal approximation to part (b) close to your findings? How close?

5.2.5 (a) Find the mean and standard deviation of the distribution in Problem 4.7.2.

(b) Determine the per cent of scores that are within 2σ of the mean.

(c) Is the normal approximation to part (b) close to your findings? How close?

In the area of inferential statistics, where judgments about populations are made from examining samples taken from the populations, there are two approaches to decision making: *hypothesis testing* and *estimation*. When we have reason to believe that an event will occur with a certain probability, we create a *hypothesis*, an assumption based upon the belief. Then we choose a statistical test and determine, with some degree of certainty, if our hypothesis should be accepted or rejected. For example, if our observations lead us to believe that one out of three smokers has quit smoking, a hypothesis could be made that one out of three smokers has stopped smoking. We would then have to test this hypothesis. We would do this by choosing a random sample of people who have smoked at some time and by then deciding if the proportion of smokers in the sample who quit smoking was close to what we had hypothesized. Would you reject the hypothesis if we found 10 of 30 who stopped? 9 of 30? 8 of 30? Inferential statistics aids us in deciding how close is close enough. However, when we have no information to make an assumption, our decision making is then aided by a method known as *estimation*. In this case we create an estimate by determining an interval, based upon the sample data, which we believe, with some degree of confidence, contains the population mean or proportion. For example, if we have no knowledge as to the per cent of smokers who quit, a sample could be taken from the population of people who have ever smoked and a confidence interval can be created as an estimation of the population who had quit smoking. In this chapter we shall examine hypothesis testing and estimation with respect to binomial probability situations.

5.3 Hypothesis Testing

Suppose that a couple rents a cabin for a weekend. As night falls, along with temperature, the cabin cools as the fire fades. The couple decide that one of them must go out to get some more wood for the fire. Bob suggests, "Flip my lucky coin to see who goes." Isabeth notices that Bob is smiling and thinks that the coin may not be fair. Bob contends that the coin is fair; that is, the *hypothesis* is that the coin is fair: $P(\text{heads}) = P(\text{tails}) = .50$. The *alternative hypothesis* is that the coin is not fair: $P(\text{heads}) \neq .50$. While Bob is getting dinner ready, Isabeth decides to test the coin. The decision rule she decides is to reject the coin as fair if the number of heads in five flips is 0 or 5. The *rejection* or *critical region* is $\{0, 5\}$. The *acceptance region* is $\{1, 2, 3, 4\}$. Isabeth flips the coin five times and counts 1 head. The hypothesis has been tested with sample data and is not rejected and Isabeth is satisfied that the coin is fair.

If the coin had landed all heads or all tails, Isabeth would have rejected the hypothesis that the coin was fair. However, a fair coin could, in fact, fall five times all heads or all tails and still be fair. Whenever we make judgments about populations based upon our sample findings (e.g., flipping the coin five times), there can always be an error in judgment because the sample, by pure chance, might demonstrate an outcome that a different sample might not. The sampling data, by chance, could give a false impression of the population. As you might expect, when a true hypothesis is rejected, an error is made. Rejecting a true hypothesis is known as a *type I error*. The probability of committing a type I error is known as the *level of significance* and is written as α (alpha). We shall be concerned with examining this error and ways to minimize it, along with a second type of error which we shall discuss shortly.

In the coin-flipping cabin problem, as in every test of a hypothesis, the type I error is present. If the coin had landed 0 or 5 heads, Isabeth would have rejected the coin, when actually it could have been a fair coin. The probability of this error, the level of significance, is found by using the binomial probability distribution. We then have

$$\alpha = P(H = 0) + P(H = 5)$$

$$= \binom{5}{0}\left(\frac{1}{2}\right)^0\left(\frac{1}{2}\right)^5 + \binom{5}{5}\left(\frac{1}{2}\right)^5\left(\frac{1}{2}\right)^0 = .062$$

Over 6 per cent of the time, a fair coin will fall all heads or all tails in five trials. If Isabeth had decided as a decision rule to reject the coin if 0, 1, 4, or 5 heads landed face up, then

$$\alpha = P(\text{type I error})$$

$$= \binom{5}{0}\left(\frac{1}{2}\right)^0\left(\frac{1}{2}\right)^5 + \binom{5}{1}\left(\frac{1}{2}\right)^1\left(\frac{1}{2}\right)^5 + \binom{5}{4}\left(\frac{1}{2}\right)^4\left(\frac{1}{2}\right)^1 + \binom{5}{5}\left(\frac{1}{2}\right)^5\left(\frac{1}{2}\right)^0$$

$$= .376$$

In this situation Isabeth would have rejected a fair coin more than 37 per cent of the time. This critical region {0, 1, 4, 5} is so large that over the long run, almost 38 of 100 coins will be rejected using this decision rule.

Consider the following situation, in which a person who claims to have extrasensory perception is to be tested. Twenty slips of paper, half of which contain an *x*, are folded and mixed thoroughly. If there is no reason to believe that this person has ESP, and is just as likely to pick an *x* as not,·the hypothesis would be $P(\text{picking an } x) = .50$, expressing an even chance. We call this hypothesis the *null hypothesis* (namely, the person does not have ESP) as opposed to the alternative hypothesis (namely, the person has ESP). The null hypothesis is usually written as H_0, while the alternative hypothesis is written as H_1 or H_A—we will use H_A. For the ESP problem, we are testing the null hypothesis, $H_0 : P(x) = .50$ against the alternative hypothesis, $H_A : P(x) > .50$. The alternative hypothesis tells us the direction of the critical region. In this situation, if the number of correct responses is significantly more than 50 per cent, we shall reject the null hypothesis that the person does not have ESP.

If we choose $n = 10$ as the number of slips of paper to be picked, we must decide upon the number of "x"'s picked that will convince us statistically that the person has ESP. If seven were picked, would that be convincing? Eight? Nine? Our decision rule is of the form, "Reject the null hypothesis if the number of successes, x is 'too many' ("too many" must be expressed numerically)." From Table II in Appendix B, we find that for $n = 10$ and $p = .50$, $P(x = 8) = .044$, $P(x = 9) = .010$, and $P(x = 10) = .001$. If the person did not have ESP, that is, if $P(x) = .50$ (just guessing), the chance that someone would pick 8 of 10 (by guessing) is less than 5 times in 100; 9 of 10 is 1 in 100, and 10 of 10 is 1 in 1,000. In determining the members of the rejection or the critical region, we want to include those possibilities that would rarely occur if the null hypothesis were true. We must be aware that the more values we include in the critical region, the greater the chance that we will reject a true hypothesis. On the other hand, if we create a small critical region, the chance is greater that we will accept a false hypothesis. It is by common agreement that the critical region should usually contain between 1 and 10 per cent of the total possibilities. The common choices of α are .01, .05, or .10—with $\alpha = .05$ used most often in the situations that we shall examine. So we want to include those values in the critical region whose probability of occurrence is at most 5 per cent. By doing this we guarantee that we will not reject a true hypothesis more than 5 per cent of the time, over the long run.

In the ESP problem, then, the critical region contains 9 and 10 successes. That is, if in 10 picks the person correctly finds either 9 or 10 "x"'s, we shall reject the hypothesis and say that the individual has ESP. The level of significance, or the probability of rejecting a good hypothesis, is found by calculating $P(x = 9) + P(x = 10) = .010 + .001 = .011$. A person who does not have ESP will pick 9 or 10 correct, over the long run, in only 11 times out of 1,000. There is a very small chance we would reject the null hypothesis if the person did not have ESP.

At this point, after the null and alternative hypotheses have been created, the level of significance and the critical region determined, the test sample is chosen. For example, if the person in the ESP experiment picks 8 x's, the hypothesis is not rejected. Only if 9 or 10 x's would have been picked would the null hypothesis of guessing be rejected. If the critical region had been $\{8, 9, 10\}$, the null hypothesis would have been rejected and the claim of ESP accepted. If this were the case, the $P(\text{type I error}) = .055$—which is very close to the .05 level of significance set originally. Would you have felt that if 8 x's were picked they would not be enough to reject the hypothesis? You can see that there may be a very thin line between accepting or rejecting the null hypothesis.

Hypothesis tests, such as in the coin-tossing cabin problem, are referred to as *two-tailed* or *two-sided* tests when the critical region contains both ends of the set of possible outcomes (e.g., {0, 5}). If the number of heads was too few or too many (two tails), the null hypothesis was rejected. The critical region is determined first by the alternative hypothesis and then by the level of significance. The alternative hypothesis determines the direction of the values in the critical region, and the level of significance determines the number of values. In the ESP situation, the test is called a *one-tailed test to the right*, as $H_A : P > .50$. The test determines a critical region with values on the extreme right of the set of possibilities ({9, 10}). Thus, if too many correct guesses were made, the null hypothesis of guessing would be rejected. The value of alpha (.05) would be represented by an area that would contain 5 per cent of the total area and would lie on the right side of the distribution. (Can you describe a one-tailed test to the left?) If $\alpha \leq .05$, then each tail of a two-tailed test would contain at most .025 so that that the total area of the rejection region would be at most 5 per cent. The shaded portions of Figures 5.3.1 and 5.3.2 represent the critical regions of a one-tailed test to the right and a two-tailed test, respectively. In this binomial probability distribution, $n = 8$, $p = .40$, and $\alpha \leq .05$.

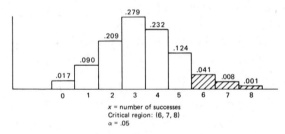

Figure 5.3.1 One-Tailed Test to the Right ($H_A : P > .40$)

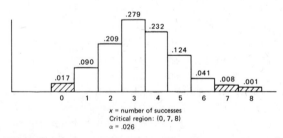

Figure 5.3.2 Two-Tailed Test ($H_A : P < .40$ or $P > .40$)

As an example to illustrate the decision making required in deciding to use a one-tailed or two-tailed test, consider the situation where a handwriting expert declares that he is able to determine if an individual has committed a crime in the past from looking at his signature. Those who doubt him have devised the following test. They will show him 22 pairs of signatures, where one of each pair is, in fact, the signature of a person with a criminal record, and the other the signature of a person who does not have a criminal record. Each trial will consist of the expert choosing that individual of the pair he believes to have a criminal record. Those who think he is just guessing want to use as the null hypothesis, $H_0 : P = .50$. What is their alternative hypothesis? Is it that he does see something $(P > .50)$? On the other hand, could the possibility exist that the expert does, in fact, see something in the signatures but that he is interpreting it just the opposite—that is, he usually picks the noncriminal? Granting this possibility that he may be finding something but interpreting it incorrectly, the alternative hypothesis, $H_A : P < .50$ or $P > .50$ is created. Thus the test will be two-tailed, and the rejection region will contain values from both ends of the distribution.

To determine the left critical region, we look under the $n = 22$ column at the smaller values of x where $p = .50$. We find that if $x = 0, 1, 2, 3, 4$, or 5, than $P(x = 0) + P(x = 1) + P(x = 2) + P(x = 3) + P(x = 4) + P(x = 5) = .008$. If "$x = 6$" would also be included, the left critical region would be greater than .025, so "$x = 6$" must be omitted. The left critical region is $\{0, 1, 2, 3, 4, 5\}$. The right critical region is $\{17, 18, 19, 20, 21, 22\}$, as $P(x = 17$ or $x = 18$ or $\cdots x = 22) = .008$. Thus the two-tailed critical region is $\{0, 1, 2, 3, 4, 5, 17, 18, 19, 20, 21, 22\}$, with a level of significance of $.008 + .008 = .016$. The decision rule is: Reject the null hypothesis that he is guessing if the expert correctly finds at most 5 (he sees something but is interpreting it incorrectly) or at least 17 (he sees something). As $\alpha = .016$, there are 16 chances out of 1,000 that although he does not see anything, he will by chance alone pick a number of correct signatures that falls in the critical region. The hypothesis testers are willing to take this chance of committing a type I error.

Suppose that the expert guesses 16 correctly; then the hypothesis is not rejected. Some of the others argue that only a one-tailed test to the right should have been used where the alternative hypothesis should be, $H_A : P > .50$. In this case the critical region is $\{16, 17, 18, 19, 20, 21, 22\}$, with $\alpha = .026$. Now the hypothesis that he is guessing is rejected as 16 falls in the rejection region. You can see how crucial the formulation of the null and alternative hypotheses are. They should be chosen before the sample is taken. An additional complication is the consideration of the acceptable level of significance. While we have used .05, in other situations, values as low as .001 or as large as .20 may be used. For example, a pharmaceutical manufacturer who plans to test a new drug has to consider the availability of other drugs that treat the same condition. If the drug they are considering marketing is very similar to many others on the market, sales will be quite limited and the project may be a financial disaster. So, if there are many available, they want to demonstrate that their drug is very effective. Should they set a low or a high level of significance? If they set a low level of significance, the critical region will be relatively small, and the probability of not rejecting this hypothesis is large. However, by setting a high level of significance, the critical region will be relatively large, and in this case if the test does not lead to rejecting the hypothesis, they know that their drug is very effective and might sell well. While the study of statistics provides tools for decision making, it is up to the person who uses them to make the ultimate decisions.

The general form of a hypothesis test can be stated as follows:

1. Create the null and alternative hypotheses, H_0 and H_A.
2. State the level of significance and the sample size.
3. Determine a decision rule based upon the alternative hypothesis and the level of significance. The form of the decision rule is: "Reject the null hypothesis if the number of successes falls in the critical region."
4. Test the null hypothesis with sample data.
5. State if the null hypothesis is rejected or not rejected, depending upon the sample data findings and the critical region.

The following hypothesis will be tested using the general form: A politician states that at least 30 per cent of his constituents own a car and a television set. Test his hypothesis using $\alpha \le .05$ and a one-tailed test to the left. (Do you see why the alternative hypothesis was to the left?) A random sample of $n = 20$ people in his area determines that three people own a car and a television set.

1. $H_0 : P = .30$; $H_A : P < .30$.
2. $\alpha = .05$; $n = 20$.
3. Reject the null hypothesis if the number of people who own a car and a television set are at most 2. [Since $P(x = 0) + P(x = 1) + P(x = 2) = .0354 < .05$.]
4. The sample value 3 does not fall in the critical region $\{0, 1, 2\}$.
5. Retain the null hypothesis; it cannot be rejected.

Note that if $\alpha = .10$, the critical region would be $\{0, 1, 2\}$, since $P(x = 0) + P(x = 1) + P(x = 2) = .04 < .10$. And the null hypothesis would not be rejected, since the sample value 3 does not fall in the critical region.

By setting the alternative hypothesis to the left $(P < .30)$, rejecting the null hypothesis would have meant that, in fact, too few people own a car and a television set, and then the politician's belief would have to be reconsidered. If we set the alternative hypothesis to the right $(P > .30)$, rejecting the null hypothesis would have meant that too many people own both items. Since the politician hypothesized that at least 30 per cent of the people, that is, 30 per cent or more, own both items, his hypothesis should not be rejected if too many people own both items.

P R O B L E M S

5.3.1 If the drug manufacturer set the null hypothesis at $P = .70$ (70 per cent cure), should the alternative hypothesis be a one-sided test to the right or left? Why?

5.3.2 If a hypothesis of $P = .60$ has been rejected by a one-sided test to the right, what can be said about the hypothesis $P = .50$ using the same critical region? Why? If $P = .60$ had been rejected by a one-sided test to the left, what could then be said of $P = .50$?

5.3.3 Some people are considering purchasing a clothing store. The seller claims that at least 70 per cent of his sales yields 100 per cent profit. (Note that the alternative hypothesis is $P < .70$.) The potential buyers observe the next five sales and find that two sales yielded 100 per cent profit. At a level of significance of .05 is the hypothesis rejected?

5.3.4 In Problem 5.3.3 it was seen that although only 40 per cent (2 of 5) of the sales yielded 100 per cent profit, the hypothesis that $P \geq .70$ could not be rejected. Consider the same hypothesis but this time where $n = 25$ purchases are observed. Can a sample average of 40 per cent still be enough not to reject the null hypothesis? What must be the minimum number of purchases out of 25 that must yield 100 per cent profit in order for the null hypothesis not be rejected?

5.3.5 Test the null hypothesis, $H_0 : P = .60$, versus $H_A : P < .60$, where $n = 20$ and $x = 8$ (sample average $= \frac{8}{20} = .40$) when (a) $\alpha = .05$ and (b) $\alpha = .10$.

5.3.6 Test the null hypothesis, $H_0 : P = .30$ versus $H_A : P < .30$ or $P > .30$, where $n = 25$ and $x = 11$, when (a) $\alpha \leq .05$ and (b) $\alpha \leq .10$.

5.3.7 The possible outcomes in a binomial distribution of n trials can be separated into a critical region and an acceptance region (those values not in the rejection region). What generalization could you make about the relationship of α and the size of the acceptance region? (Problems 5.3.5 and 5.3.6 might be helpful.)

5.3.8 The politician's claim $(P \geq .30)$ cannot be rejected when, in fact, there was exactly 10 per cent $(\frac{2}{20})$ of the random sample satisfied his claim. How do you explain this?

We have seen that by setting a small value for α, a relatively small critical region will be created. That is, the chance that we shall reject a good hypothesis is small. We could, in fact, make the chance of rejecting a good hypothesis be 0.0. In this case, we would never have a critical region and never reject any hypotheses. There would be no decisions to be made. For if we never rejected any hypothesis, then we would have to accept any hypothesis—whether it was true or false. The chance of accepting a false hypothesis, β (beta), which is the probability of committing a type II error would be certainty. For example, prior to open-enrollment programs in college, the rejection region for admission to college was set quite large, so the chance of rejecting a "good" student was large, while the chance of accepting a "poor" student was small. The philosophy of the open-enrollment program is that to reject a "good" student is the greater evil. The belief is that while the chance of accepting a "poor" student may be initially large, this error can be corrected for by examining the student's work after enrollment. Another example that demonstrates how the size of α reflects the particular situation is in the medical profession:

> In ascertaining efficacy, the pharmaceutical manufacturer engaged in testing a compound which may have a wide therapeutic use would set α at some small value, i.e., 1% (.01), thereby demanding good evidence that a compound is effective before he claims it to be so. If, however, the drug under test has a potential use which is quite limited or other effective drugs are available for those conditions in which the test compound would be used, then α could be set at some larger value, i.e., 10% to 20% (.10 to .20).*

The relationship of α, β, and n are examined in the next set of test hypotheses situations.

EXAMPLE 1

1. $H_0:P=.70; H_A:P<.70$.
2. $\alpha \le .05; n=5$.
3. Decision rule: Reject the null hypothesis if the number of successes, x, is at most 1; that is the critical region is $\{0, 1\}$, since $P(x=0)+P(x=1)=.002+.028=.03=\alpha$. And the acceptance region is $\{2, 3, 4, 5\}$.

Suppose that $P \ne .70$, but that, in fact, $P=.60$. What is the chance of accepting the false $(P=.70)$ hypothesis? If a sample of $n=5$ is taken and two or more successes are found, the (false) hypothesis is not rejected. To find β, P(type II error), we determine the chance that at least two successes could occur when in fact $P=.60$ and not .70; that is, $\beta = P(x=2$ when $p=.60)+P(x=3$ when $p=.60)+P(x=4$ when $p=.60)+P(x=5$ when $p=.60)$;

$$\beta = .230+.346+.259+.078=.913$$

---

* B. A. Barron and S. C. Bukantz, "The Evaluation of New Drugs," *Archives of Internal Medicine*, Vol. 119, June 1967, pp. 547–556.

Thus there is more than a 91 per cent chance of accepting a false $(P = .70)$ hypothesis when, in fact, the true hypothesis is $P = .60$ and $n = 5$.

What is the chance of accepting $P = .70$ when, in fact, $P = .30$? We want to find the chance of obtaining at least two successes when the probability of success is .30; that is, $\beta = P(x = 2$ when $p = .30) + \cdots + P(x = 5$ when $p = .30)$;

$$\beta = .309 + .132 + .028 + .002 = .471$$

When $P = .10$ [notice that $P(x = 4$ when $p = .10)$ and $P(x = 5$ when $p = .10)$ is very close to 0—does this make sense?]

$$\beta = .073 + .008 = .081$$

The reason β has decreased considerably from .931 to .081 is that the true hypothesis has moved farther away from the false hypothesis; that is, $P = .60$ (true) versus $P = .70$ (false) to $P = .10$ (true) versus $P = .70$ (false). The false hypothesis is easier to spot the farther away from the true hypothesis it is. An additional complication is the small sample size; it does not provide enough data for the true hypothesis to show itself clearly. The effect of sample size on β can be seen by comparing this situation to the next problem where $n = 20$ instead of $n = 5$.

EXAMPLE 2

1. $H_0 : P = .70$; $H_A : P < .70$.
2. $\alpha \leq .05$; $n = 20$.
3. Decision rule: Reject the null hypothesis if the number of successes, x, is at most 10; that is, the critical region is $\{0, 1, \ldots, 10\}$, since $P(x = 0) + P(x = 1) + \cdots + P(x = 10) = .048 = \alpha$.

Again, suppose that $P \neq .70$, but that, in fact, $P = .60$. What is the chance of accepting the false hypothesis now that n has increased from 5 to 20? When 11 or more successes are found in the sample of 20, the hypothesis is not rejected. So β is the chance of accepting the false hypothesis that occurs if 11 or more successes are found. At $P = .60$,

$$\beta = .160 + .180 + .166 + \cdots + .035 + .012 + .003 = .755$$

When $P = .30$?,

$$\beta = .012 + .004 + .001 = .017$$

When $P = .10$?,

$$\beta = .000 \text{ (to three places)}$$

Thus, by taking a larger sample, the chance of accepting the false hypotheses decrease appreciably. When $n = 5$, we would accept the false hypothesis $P = .70$ where the true hypothesis $P = .30$ is in 471 cases out of 1,000. But when $n = 20$, we would accept $P = .70$, where $P = .30$ is true in only 17 cases out of 1,000. The more we know, the less the chance of our being fooled by a false claim.

EXAMPLE 3

1. $H_0: P = .70$; $H_A: P < .70$.
2. $\alpha \le .20$; $n = 5$. (As α has been allowed to take a much larger value, the size of the critical region will increase substantially.)
3. Decision rule: Reject the null hypothesis if the number of successes, x, is at most 2; that is, the critical region is $\{0, 1, 2\}$. $P(x = 0) + P(x = 1) + P(x = 2) = .162 = \alpha$.

If $P = .60$, the chance of accepting the false hypothesis $P = .70$ is now found by determining the chance of obtaining three, four, or five successes when $P = .60$:

$$\beta = .346 + .259 + .078 = .683$$

When $P = .30$?,

$$\beta = .132 + .028 + .002 = .162$$

When $P = .10$?,

$$\beta = .008$$

Notice that in comparison to the case where $n = 5$ and $\alpha \le .05$, the β values, where $n = 5$ and $\alpha \le .20$, are considerably smaller. This is reasonable, since by creating a larger critical region there is a greater chance of rejecting a true hypothesis and thus a smaller chance of accepting a false hypothesis. To have both a small chance of rejecting a true hypothesis and a small chance of accepting a false hypothesis at the same time, the size of the sample must be large. Table 5.3.1 contains the comparisons for α, β, and n when H_0: $P = .70$ and H_A: $P = .30$.

Table 5.3.1. Relationship of α, β, and n when
$H_0: P = .70$ and $H_A: P = .30$

n	5	5	20
α	.03	.162	.048
β	.471	.162	.017

PROBLEMS

5.3.9 Determine β at $P = .30$ and $P = .50$ for Problem 5.2.3.

5.3.10 Determine β at $P = .30$ and $P = .50$ for Problem 5.2.4.

5.3.11 Can you present a good argument why $\alpha = \beta$ in the middle column of Table 5.3.1?
(HINT. Reexamine Example 3.)

5.4 Estimation

As we mentioned earlier, when we do not have any or enough information to help us create a hypothesis, we collect sample data and, based upon the findings, we can make an estimate of the population characteristics. Since we are only taking a sample (we hope it is random) and not the entire population data, there is a chance that the sample data do not exactly reflect the population. What we can do is create confidence intervals based on the sample findings that state with some degree of certainty the characteristics of the population.

When we collect a sample of size n and determine the number of favorable responses, x, to some questions, we can calculate the *sample proportion* of favorable responses, $\bar{p} = \dfrac{x}{n}$. Then we can determine lower and upper bounds of an interval around the sample proportion in which we can be 95 per cent confident that the population proportion, p, lies. This interval estimate is known as the 95 per cent *confidence interval*. It is important to note that we are not saying that p, the population proportion, lies in the interval with probability .95—because p is either in or not in the interval. We are saying that this confidence interval determined by a particular sample proportion $\bar{p}$ has a 95 per cent chance of containing p. The following example will demonstrate the relationship between $\bar{p}$ and p.

Consider the binomial situation where a sample of $n = 200$ adults includes 28 adults who are out of work. Therefore, the sample proportion, $p = \dfrac{x}{n} = \tfrac{28}{200} = .14$—14 per cent of the sample are out of work. From this piece of information we want to determine a confidence interval that we can be 95 per cent confident contains the population proportion, p. The questions we want to answer are: (a) What is the 95 per cent confidence interval for p, the population proportion of people out of work? and (b) What is the accuracy of $\bar{p}$ as an estimate of p?

(a) We know that 95 per cent of the scores in a normal distribution fall within 1.96 standard deviations of the mean. Since n is large, we can approximate quite well the binomial distribution with the normal distribution. By standardizing the binomial distribution mean and standard deviation so that $\mu = 0$ and $\sigma = 1$, we can determine binomial probabilities of compound events rather easily. To standardize a set of scores, recall that each x is converted to a z value, where $z = \dfrac{x - \mu}{\sigma}$. For the binomial distribution we have that $z = \dfrac{x - np}{\sqrt{npq}}$. As we are interested in making estimates about the population proportion p and not np, we divide the numerator and denominator of the fraction that z equals by n and obtain:

$$z = \frac{x/n - p}{\sqrt{pq/n}}$$

Notice that if $p = \dfrac{x}{n}$, that is, if the population proportion is equal to the sample proportion, the value of z is found to be zero—the sample proportion would be 0 standard deviations from the population proportion.

From the formula just derived from z, we have the sample proportion, $\dfrac{x}{n}$, and the standard deviation $\sqrt{\dfrac{pq}{n}}$. The sample proportion we found to be .14. The standard deviation cannot be determined exactly since p, the population proportion, is unknown. We use $\bar{p}$ as an estimate of p, and so $q = 1 - p = 1 - .14 = .86$. The standard deviation is then $\sqrt{(.14)(.86)/200} = .025$. So the interval that is 1.96 standard deviations around the sample proportion is

$$\left(\bar{p} - 1.96\sqrt{\frac{pq}{n}}, \bar{p} + 1.96\sqrt{\frac{pq}{n}}\right) = (.14 - 1.96(.025), .14 + 1.96(.025))$$

$$= (.091, .189)$$

We have estimated that $.091 < p < .189$, based upon $\bar{p} = .14$, n large, and the choice of a 95 per cent confidence interval.

There is a 5 per cent chance that, owing to sampling variation, this confidence interval does not contain the population proportion. We can increase our confidence to 99.7 per cent if we use 3 standard deviations in place of 1.96. We would then have

$$\bar{p} - 3\sqrt{\frac{pq}{n}} < p < \bar{p} + 3\sqrt{\frac{pq}{n}}$$

and substituting for $\bar{p}$ and $\dfrac{pq}{n}$, we have

$$.14 - 3(.025) < p < .14 + 3(.025)$$

$$.065 < p < .215$$

Notice that this interval is larger than the 95 per cent confidence interval, for we have included all the scores that contribute 99.7 per cent of the distribution based on the sample proportion $\bar{p}$. Before, we could be 95 per cent confident that p is greater than .09 and less than .189. Now we can be 99.7 per cent confident that p is greater than .065 and less than .215.

(b) In answering (a), we have determined the accuracy of $\bar{p}$ as an estimate of p. The value of

$$1.96\sqrt{\frac{pq}{n}} = 1.96(.025) = .049$$

tells us that $\bar{p}$ will not differ from p by more than .049 with 95 per cent confidence. However, there is a 5 per cent chance that p and $\bar{p}$ differ by more than .049.

We can ask, in addition: What should the size of the sample n be so that the difference between $\bar{p}$ and p is at most .03, and still be 95 per cent confident in our findings? Since $\bar{p} = .14$ and $\bar{p}$ is the estimate of p, we have that

$$1.96\sqrt{\frac{pq}{n}} = 1.96\sqrt{\frac{(.14)(.86)}{n}} = .03$$

We need to solve this equation for n:
1. Divide both sides by 1.96:

$$\sqrt{\frac{(.14)(.86)}{n}} = \frac{.03}{1.96} = .015$$

2. Square each side:

$$\frac{(.14)(.86)}{n} = .0002$$

3. Multiply both sides by n: $(.14)(.96) = .0002n$.
4. Simplify: $.1204 = .0002n$.
5. Divide by .0002: $602 = n$.

Since $\bar{p}$ was determined by taking a sample of $n = 200$, we would have to take an additional sample of 402 to guarantee with 95 per cent confidence that $\bar{p}$ and p do not differ by more than .03.

PROBLEM

5.4.1 Suppose that a sample of $n = 300$ people were randomly chosen and 36 people stated that they have cut down on the number of foods they buy which contain chemical additives. We can use the binomial probability model since n is known, the person has either cut down or not cut down, the trials are independent, and we assume that $\bar{p} = \dfrac{x}{n}$ is constant. Determine:
  (a) The accuracy of the sample proportion as an estimate for p for $z = 1.96$.
  (b) A 95 per cent confidence interval for p.
  (c) A 99.7 per cent confidence interval for p.
  (d) The size of the sample necessary if one is to be 95 per cent confident that the difference between p and $\bar{p}$ is at most .02.

If the normal distribution was not appropriate as an approximation (*n* was small) to the binomial distribution, then, in order to determine confidence intervals, we would have to find, after knowing $\bar{p}$ and *n*, the critical region. When we separate the critical region from the set of possible outcomes, we would be left with the acceptance region or the confidence interval. For example, if the set of possible outcomes was {0, 1, . . . , 10} and the critical region was {0, 1, 10}, then the acceptance region would be {2, 3, . . . , 9}. However, this would require our finding *P(x)* for each outcome for each sample size. Fortunately, the information has all been put in a readily available form (Figure 5.4.1). For example, if $\bar{p} = .40$ and *n* = 20,

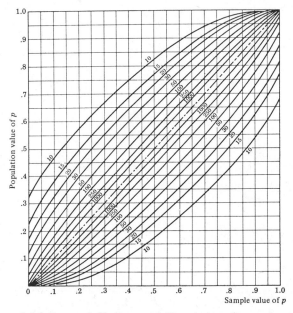

Figure **5.4.1** Interval Estimate of Population Proportion *P* with Confidence Coefficient .95 for *N* = 10, 15, 20, 30, 50, 100, 250, and 1,000. (From *Biometrika*, by C. J. Clopper and E. S. Pearson, 1934. Reprinted by permission of the Biometrika Trustees.)

then by looking along the vertical line of $\bar{p} = .40$ we find that this line intersects the *n* = 20 bottom curve at *p* = .18 (approximately) and the *n* = 20 top curve at *p* = .65 (approximately). Thus the 95 per cent confidence interval for $\bar{p} = .40$ and *n* = 20 is .18 < *p* < .65. The values of *p* above the top curve and below the bottom curve fall outside the acceptance or confidence interval.

If we take a large sample, notice what happens to the width of the interval without changing the level of confidence. For example, if we had found $\bar{p} = .40$, in a sample of *n* = 1,000, the 95 per cent confidence interval would be $.38 < p < .43$. Notice that this interval is a much better estimate than $.18 < p < .65$ when $n = 20$. You can get a good idea why national opinion polls usually obtain about 1,550 people's opinions and are confident that the sample findings reflect the total population. Of course, the way the sample is taken is crucial. The 1,550 people must be representative of the entire population.

Figure 5.4.1 can also be used in hypothesis testing. By drawing a horizontal line through the assumed value of p (population value), one can determine the sample proportion interval that is acceptable at a level of significance of .05. For example, suppose the claim is that $H_0: P = .65$ and $H_A: P < .65$ or $P > .65$ when $n = 50$. The horizontal line through $P = .65$ intersects the $n = 50$ top curve at $\bar{p} = .50$ and the bottom curve (at the right) at $\bar{p} = .75$. Thus, for the null hypothesis not to be rejected, the sample proportion must fall in the interval $.50 < \bar{p} < .75$. As $n = 50$, the number of successes, x, needed not to reject the hypothesis would be $.50(50) < x < (.75)(50) = 25 < x < 38$; in other words, the acceptance region is $\{26, 27, \ldots, 37\}$. Accordingly, the critical region would be

$$\{0, 1, \ldots, 25, 38, \ldots, 50\}$$

Consider the following situation. A manufacturer claims that at least 70 per cent of her paintbrushes are durable. A retailer decides to test the brushes by taking a random sample of $n = 20$ and putting them to a durability test. If the manufacturer's claim is not rejected, the retailer will buy a large shipment of brushes. The null hypothesis is $H_0: P \geq .70$ is set up as $H_0: P = .70$ and the alternative hypothesis $H_A: P < .70$. In this way, if the null hypothesis is rejected, the statement "that at least 70 per cent of the brushes are durable" is rejected. The retailer is willing to accept a 2.5 per cent type I error. The question is: What is the acceptable range for the sample proportion that is required not to reject the null hypothesis? If we draw a horizontal line through $P = .70$, we find that it cuts the top (left) band for $n = 20$ at approximately $\bar{p} = .46$, and the bottom (right) band at $\bar{p} = .92$. However, the test is one-sided to the left. Therefore, $\bar{p} = .92$ is disregarded, for surely if $\bar{p} = .92$, the null hypothesis would not be rejected. What must be satisfied is that $\bar{p} > .46$. If it is not, the hypothesis is rejected. If $\bar{p} > .46$, in a sample of $n = 20$ there must be $.46(20) = 9.2$, or 10 acceptable brushes. Notice that if only 50 per cent (10 of 20) are durable, the claim of $P = .70$ cannot be rejected. This problem is due in large part to the small sample size (see Problem 5.4.5).

P R O B L E M S

5.4.2 Determine the 95 per cent confidence interval for the population proportion if:
   (a) $n = 10$ and $x = 5$ (that is, $\bar{p} = x/n$).

(b) $n = 50$ and $x = 25$.

(c) $n = 1,000$ and $x = 500$.

5.4.3 If $H_0 : P = 60$ and $H_A : P < .60$ or $P > .60$ and $\alpha = .05$, test the hypothesis if a sample of $n = 30$ yielded 13 successes. What is the critical region?

5.4.4 What does the straight line between the $n = 1,000$ curves represent?

5.4.5 If a sample of $n = 100$ paintbrushes was tested with $\alpha = .025$, what would the sample proportion have to be in order not to reject the null hypothesis that $P \geq .70$? How many acceptable brushes must there be?

5.5 Summary

In this chapter we examined the properties and applications of the binomial probability distribution. We saw that whenever a sample is taken from a population where the probability of success is assumed constant, the trials are independent, and the outcomes can be categorized in two ways, the binomial probability model can be used to determine the probability of compound events. The binomial model was seen to be a valuable tool in decision making for situations where hypotheses were tested and confidence intervals were estimated.

Hypothesis testing involved testing with sample findings assumptions made about population proportions or percentages of success. Each null hypothesis was contrasted to an alternative hypothesis. The level of significance and the alternative hypothesis determined the critical region. If upon testing the hypothesis with sample data the number of outcomes fell in the critical region, the null hypothesis was rejected. We developed a general format for testing hypotheses:

1. Create the null and alternative hypotheses, H_0 and H_A.
2. State the level of significance and the sample size.
3. Determine a decision rule based upon the alternative hypothesis and the level of significance. The form of the decision rule is: Reject the null hypothesis if the number of successes fall in the critical region.
4. Test the null hypothesis with the sample findings.
5. Accept or reject the null hypothesis depending upon the sample data and the critical region.

Two errors, type I and type II, are always part of hypothesis testing. A type I error is the error made in rejecting a true hypothesis. A type II error is the error made in retaining a false hypothesis. Since the size of the rejection region is determined by the level of significance, the larger the level of significance, the larger the critical region and the greater the chance that a true hypothesis will be rejected. By choosing a small level of significance, the chance of rejecting a true hypothesis will be small, but the chance of accepting a false hypothesis becomes large. In order to minimize both types of errors, the same size, n, must be increased.

When one has too little information to make an assumption, an estimate of a population proportion can be made by using the sample proportion. This point estimate is expanded into an interval estimate by choosing a level of confidence that is converted into units of standard deviations from the sample proportion. The normal distribution was used as an approximation to the binomial distribution. By doing this, we can use the information that 1.96 standard deviations around the mean includes 95 per cent of the total frequency of scores, and so a 95 per cent confidence interval can be created in which the population proportion is expected to lie. By choosing n appropriately, one can determine the accuracy of how close $\bar{p}$, the sample proportion, is to p, the population proportion. Since the normal distribution is invaluable in estimating confidence intervals for the binomial distribution and is widely used in statistics, Chapter 6 is devoted to the normal distribution.

5.6 Review of Formulas

  I. Binomial distribution
    A. The probability of obtaining exactly k successes in n trials is

$$\binom{n}{k}p^k q^{n-k} = P(x) = \frac{n!\,p^k q^{n-k}}{k!\,(n-k)!}$$

       where p is the probability of success and q is the probability of failure.
    B. The probability of obtaining at least k successes in n trials is

$$P(x \geq k) = \sum_{x=k}^{n} \frac{n!\,p^x q^{n-x}}{x!\,(n-x)!}$$

    C. The mean is $\mu = np$, and the standard deviation is $\sigma = \sqrt{npq}$.
  II. Hypothesis testing
    A. Null hypothesis, H_0, is tested against an alternative hypothesis, H_A.
      1. One-tailed test: The alternative hypothesis is either "$P <$ ____" (left-tailed), or "$P >$____" (right-tailed). The critical region contains a set of values that reflect the chosen level of significance, α, and the alternative hypothesis. For a test to the left, the critical region is of the form $\{0, 1, \ldots, k\}$ such that $P(x = 0) + P(x = 1) + \cdots + P(x = k) \leq \alpha$. For a test to the right, the critical region is of the form $\{j, j+1, \ldots, n\}$ such that $P(x = j) + \cdots + P(x = n) \leq \alpha$.

2. Two-tailed test: The alternative hypothesis H_A is of the form "$P \leq$ or $P \geq$." The critical region is of the form $\{0, 1, \ldots, h, t, \ldots, n\}$ such that $P(x=0)+\cdots+P(x=h)\leq\frac{1}{2}\alpha$ and $P(x=t)+\cdots+P(x=n)\leq\frac{1}{2}\alpha$. Thus the probability of the total critical region contains the level of significance, α, determined prior to testing the hypothesis.

III. Estimation
   A. Confidence interval
      1. The 95 per cent confidence interval for the population proportion is $\left(\bar{p}-1.96\sqrt{\dfrac{pq}{n}}, \bar{p}+1.96\sqrt{\dfrac{pq}{n}}\right)$, where $\bar{p}$ is the sample proportion, n is the size of the sample, p is the population proportion approximated by $\bar{p}$, and $q=1-p$.
      2. The 99.7 per cent confidence interval for the population proportion is $\left(\bar{p}-3\sqrt{\dfrac{pq}{n}}, \bar{p}+3\sqrt{\dfrac{pq}{n}}\right)$.

By increasing the units of standard deviations from the mean from 1.96 to 3, the percentages of scores within the interval is increased from 95 to 99.7. While the level of confidence has increased from 95 to 99.7, the width of the interval has also increased.

5.7 Problem Set

5.7.1 What is the probability of obtaining exactly four tacks landing face up in 10 trials if the probability of not landing face up is .8?

5.7.2 At least four face up?

5.7.3 What is the mean and standard deviation of a binomial distribution where $p=.2$ and $n=10$?

5.7.4 If a binomial distribution has $p=.2$ and $n=10$, what per cent of the outcomes falls within 1.5 standard deviations of the mean?

5.7.5 If in an area of the city it has been determined that 30 per cent of the families have an automobile, how many families would you expect to question if you need a sample of 50 car owners?

5.7.6 If in a test of a hypothesis that $P=.60$, where $n=20$ and $\alpha=.05$, would you reject this hypothesis if $x=8$ for:
(a) A two-tailed test?
(b) A one-tailed test to the left?
(c) A one-tailed test to the right?

5.7.7 Reconsider Problem 5.7.6 if $\alpha\leq.10$.

5.7.8 A sample of 10 observations yields six successes.
  (a) Which hypothesis about the population proportion would you make?
  (b) Assuming your null hypothesis and a two-tailed alternative hypothesis, find the critical region if $\alpha \le .05$. Find β at .10 and .30 units to the left of your null hypotheses.
  (c) Reconsider parts (a) and (b) if $n = 25$ and $x = 15$.

5.7.9 If $\bar{p} = .20$, compare the 95 per cent confidence intervals using the normal distribution approximation and the chart of Figure 5.4.1 if:
  (a) $n = 10$.
  (b) $n = 50$.
  (c) $n = 1,000$.
  What do you notice about the intervals as n increases?

5.7.10 If a random sample of 100 children aged 7 or less resulted in 67 children saying they never heard of Lamont Cranston, determine:
  (a) A 95 per cent confidence interval for the population proportion.
  (b) Determine n so that the difference between $\bar{p}$ and p is at most .02 unit.

5.8 Cumulative Review

5.8.1 An automobile advertisement in a newspaper stated that all their prices of cars were decreased an average of $300. The models and the reductions were:

2-door coupe	$300
2-door sedan	$400
4-door sedan	$300
Station wagon	$200

  (a) Which average are they using?
  (b) What assumption is being made?

5.8.2 Given the histogram on p. 118, determine the following:
  (a) Mean.
  (b) Median.
  (c) Mode (modal class).
  (d) 95th percentile.
  (e) Per cent of scores within 1 standard deviation of the mean.

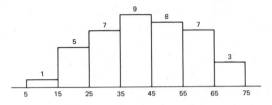

5.8.3 Suppose that one car of each of the four models was available (Problem 5.8.1).
   (a) Determine the mean, standard deviation, and range of the decreases.
   (b) Find the coding equation that transforms the set of data into $-1, 0, 0, 1$. Find the mean and standard deviation.

5.8.4 If the mean grade on a final exam was 72 and the standard deviation was 6, what scores are 2.5 standard deviations above and below the mean?

5.8.5 After a tough week at school, Letitia decided to go out Friday night. She wanted to choose a place to eat, a movie, and a late night show. She was interested in four places for dinner, five movies, and three late night shows. How many different Friday nights could she create?

5.8.6 If Letitia usually likes 75 per cent of the movies she sees, 50 per cent of the places at which she eats, and 80 per cent of the late night shows she sees, find the probability that she will like:
   (a) All three activities.
   (b) None of the activities.
   (c) At least one of the activities.

5.8.7 If Ron sees 25 movies a year and enjoys around 60 per cent of them, find the probability that he will enjoy:
   (a) All of them.
   (b) None of them.
   (c) At least 80 per cent of them.
   (d) At most 25 per cent of them.

5.8.8 Suppose that a game had the following rules and payoff. Pick a three-digit number from 000 to 999: (a) if the first digit matches the winning first digit, the person collects 6 units plus the unit bet for each unit bet; (b) if the first two digits match, the person collects 60 units plus the unit bet for each unit bet; and (c) if all three digits match, the person collects 600 units plus the unit bet for each unit bet. If Harvey bets 1 unit, what is his expectation of winning?

5.8.9 Russell just rented a music studio. He thinks something is wrong with his phone as he thinks at least 30 per cent of the calls that he receives are wrong numbers. Test his claim if in the next 20 calls he gets three wrong numbers when:
(a) $\alpha \leq .05$.
(b) $\alpha \leq .10$.
Find the β values for $P = .20$ for both parts (a) and (b).

5.8.10 What would be a 95 per cent confidence interval for the population proportion if $n = 200$ and $\bar{p} = .30$? Use both the normal approximation and the confidence interval chart (Figure 5.4.1) to determine the interval. Which interval should, in fact, be closer to the real interval?

5.8.11 Sandra is one of 15 per cent of the population who is left-handed. Find a 95 per cent confidence interval of the number of left-handed desks a school with a population of $n = 1,000$ should have.

5.8.12 Here are two other examples in which "independence" needed to be examined carefully:
a   When Dr. Benjamin Spock was on trial, his defense counsel found that of the 88 prospective jurors only 5 were women.
(b) A black civil rights worker was resisting the military draft in 1967 in South Carolina. In this state blacks comprise 34.8 per cent of the population. On the draft board in this state there were 161 people; 160 were white.
What can you say about the probabilities of these events occurring by chance?

Chapter 6
Normal Distribution

6.1 Introduction

In this chapter we shall examine at greater length the normal distribution. This distribution has applications to many different areas, such as business, education, economics, medicine, and insurance. The form of this continuous distribution is bell-shaped, symmetric, tailing off at both ends. It is determined by the mean, μ, and the standard deviation, σ, of the distribution (Figure 6.1.1). By examining Figure 6.1.1, can you guess

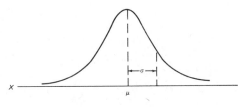

Figure 6.1.1. *The Normal Distribution*

where the median and mode of the distribution are located? Since the distribution is symmetrical, the value on the X-axis at the center of the distribution must be the median, since 50 per cent of the distribution falls above and 50 per cent below this point. As the mode is the score that occurs most frequently, it is also at the same point as the median. What about the mean? Since the mean can be thought of as the balancing point of the distribution, it, too, falls at the center of the distribution.

The fact that the normal distribution is determined by the mean and the standard deviation of the distribution tells us that by determining the mean we have found the center of the distribution, and by determining the standard deviation we have found the measure of the spread of the distribution. An interesting property of the normal distribution is that while we could draw many normal distributions with different means and standard deviations, within 1 standard deviation of the mean (from both sides) there must be very close to 68.3 per cent of the total area of the distribution; within 2 standard deviations (2σ) from both sides, there must be close to 95.5 per cent of the total area; and within 3σ of the mean (yes, from both sides) there must be around 99.7 per cent of the area of the

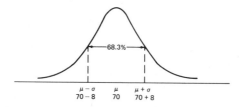

Figure 6.1.2. *Normal Distribution with* $\mu = 70$, $\sigma = 8$

distribution (Figure 6.1.2). For example, if $\mu = 70$ and $\sigma = 8$, then, *if* the distribution were normal, there would be around 68.3 per cent of the distribution between 62 and 78. What is the interval that is 2 σ's from the mean? You should find it to be (54, 86)—this interval contains approximately 95.5 per cent of the total area.

Since this relationship between percentage of area and standard deviations from the mean always holds for a normal distribution, we can use it to compare normal distributions. For instance, if two normal distributions had the same mean (e.g., $\mu = 50$) but different standard deviations (e.g., $\sigma_1 = 4$ and $\sigma_2 = 12$), we would know that the normal distribution with the smaller standard deviation would have to be "taller and thinner" than the "short and squatty" normal distribution with $\sigma_2 = 12$ (Figure 6.1.3). This is due to

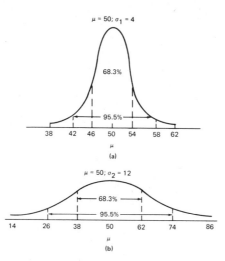

Figure 6.1.3. *Two Normal Distributions*: (*a*) $\mu = 50$, $\sigma = 4$; (*b*) $\mu = 50$, $\sigma_2 = 12$

the fact that the normal distribution with the smaller standard deviation, $\sigma_1 = 4$, contains 68.3 per cent of the area within the 46 to 54 range, while the normal distribution with $\sigma_2 = 12$ requires a range of 38 to 62 to contain 68.3 per cent of the total area.

PROBLEMS

6.1.1 (a) In Figure 6.1.2, what interval about the mean contains 99.7 per cent of the scores?

(b) Approximately what percentage of scores is contained in the interval from 70 to 78?

(c) What percentage of scores falls in the interval 54 to 78?

6.1.3 Can you think of another distribution besides the normal distribution which has the mean, median and mode at the same center point? (HINT. In plane geometry, its base angles were congruent.)

6.1.2 In Figure 6.1.3, which interval contains a greater percentage of the distributions: $(38, 62)$ when $\mu = 50$ and $\sigma = 4$, or $(14, 86)$, when $\mu = 50$ and $\sigma = 12$?

6.2 The Standardized Normal Distribution

In Chapter 2 we saw that we could compare distributions by converting sets of x-scores to sets of z-scores; that is, we standardized the scores, making them easier to compare. For example, we could standardize the two normal distributions in Figure 6.1.3 so that each would have a mean $\mu = 0$, and a standard deviation, $\sigma = 1$. By doing this, we would know right away that a value of $z = 2$ would mean that 47.75 per cent ($\frac{1}{2}$ of 95.5 per cent) of the area of the normal distribution, whatever its original mean and standard deviation was, would lie in the interval μ to $\mu + 2\sigma$ or 0 to 2 (Figure 6.2.1).

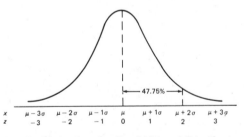

Figure 6.2.1. *Standardized Normal Distribution*

In Figure 6.1.3, where $\mu = 50$ and $\sigma = 12$, 47.75 per cent of the scores would be in the interval 50 to 74. When the x-scores are standardized, the score of $x = 50$ would convert to $z = \dfrac{x-u}{\sigma} = \dfrac{50-50}{12} = \dfrac{0}{12} = 0$, and the score of $x = 74$ would convert to $z = \dfrac{x-u}{\sigma} = \dfrac{74-50}{12} = \dfrac{24}{12} = 2$. Thus the area under the curve from 50 to 74 in the original normal distribution is equal to the area under the curve from 0 to 2 in the standardized normal distribution, since the original distribution had a mean $\mu = 50$ and a standard deviation $\sigma = 12$, and $\mu + 2\sigma = 50 + 2(12) = 74$. What standardized interval is equivalent to $(26, 74)$ of Figure 6.1.3 and so contains 95.5 per cent of the area under the curve?

In order to determine the area under the curve from $z = 0$ to $z = 1.56$ or any other z, we can use Table III in Appendix B. We look down the left-hand column until we locate "$z = 1.5$" and then across until we reach the "6" column; this value, $z = 1.56$, is associated with .4406—44.06 per cent of the total area lies between $z = 0$ and $z = 1.56$ (Figure 6.2.2). From

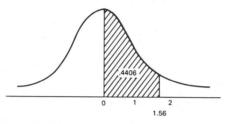

Figure 6.2.2. *Standardized Normal Distribution*

the table you should determine that if $z = 1.96$, then the area under the standardized normal distribution from $z = 0$ to $z = 1.96$ is 47.5 per cent of the total area. What is the area between $z = -1.96$ and $z = +1.96$? Since the curve is symmetric about the mean ($z = 0$), there is the same per cent under the curve from $z = -1.96$ to $z = 0$ as there is from $z = 0$ to $z = 1.96$. Thus, from -1.96 to 1.96 there is 95 per cent of the distribution. The symmetry of the distribution allows us not to have to include values for z less than zero in the table. It can always be determined by remembering that the curve is symmetric. For example, if we wanted to find the area between $z = -1.3$ and $z = 0$, we would know that this area would be the same as from $z = 0$ to $z = 1.3$: namely, .4032 (when would z be negative?).

Before we can use the normal distribution to test hypotheses and determine confidence estimates, we need to be able to determine percentages under the normal curve in many situations. For example, consider the following problem: Sales of a soft-drink company were normally distributed over the course of a year with average weekly sales of $13,000 and a standard deviation of $5,000.

1. What per cent of sales was at most $18,000? Since $\mu = 13,000$ and $\sigma = 5,000$, 18,000 is 1 standard deviation above the mean—$\mu + 1\sigma = 13,000 + 1(5,000) = 18,000$. The area under the curve in which we are interested is depicted in Figure 6.2.3. Table III tells us that between $z = 0$ and $z = 1$, there is .3413 of the total area. Since there is .5000 up to $z = 0$,

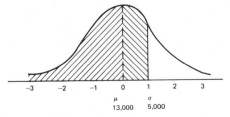

Figure 6.2.3. *Standardized Normal Distribution*

we have then that the per cent of sales at most $18,000 ($z \leq 1$) is .5000 + .3413 = 84.13 per cent. [Do you see that another interpretation of this is that $P(x \leq 18,000) = P(z \leq 1) = .8413$?]

2. What per cent of the sales were above $18,000? The area under the curve in which we are interested is found in Figure 6.2.4. Since 50 per cent

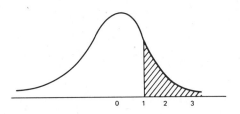

Figure 6.2.4. *Standardized Normal Distribution*

or .5000 lies under the curve from $z = 0$ to the right, then as the area from $z = 1$ to $z = 0$ contains .3413, there must be .5000 − .3413 = .1587 from $z = 1$ to the right. That is, 15.87 per cent of the sales were more than $18,000. Note that this could have been found by seeing that 84.13 per cent of the total area was contained in the region to the left of $z = 1$, so $1 − .8413 = .1587$ is the area to the right of $z = 1$.

3. What per cent of the sales fell between $12,000 and $15,000? We want to find the sum of the areas from $12,000 to $13,000, and from $13,000 to $15,000 (Figure 6.2.5). The reason that we break up this region as we did is because the mean $\mu = \$13,000$ falls in the $12,000 to $15,000 interval, and Table III determines areas from the mean to some other

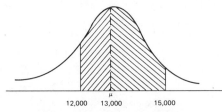

Figure 6.2.5. *Normal Distribution*

point. An $x = 12,000$ converts to $z = \dfrac{12,000-13,000}{5,000} = \dfrac{-1,000}{5,000} = -.2000$.

The area from $z = -.2$ to $z = 0$ is the same as the area from $z = 0$ to $z = .2$. Thus $.0793$ is the proportion of sales from \$12,000 to \$13,000. An $x = 15,000$ converts to $z = \dfrac{15,000-13,000}{5,000} = \dfrac{2,000}{5,000} = .4000$. The area from $z = 0$ to $z = .4$ is $.1554$ of the total area. Thus $.0793 + .1554 = 23.47$ per cent of the sales fell between \$12,000 and \$15,000.

4. What per cent of the sales was between \$14,000 and \$25,000? The area we are looking for is to the right of the mean (Figure 6.2.6). In order

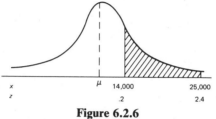

Figure 6.2.6

to find the percentage of the area in this region, we first convert the x-scores to z-scores and then use the table. An $x = 14,000$ converts to $z = \dfrac{14,000-13,000}{5,000} = \dfrac{1,000}{5,000} = .2000$, and $x = 25,000$ converts to $\dfrac{25,000-13,000}{5,000} = \dfrac{12,000}{5,000} = 2.4$. We want to find the area between $z = .2$ and $z = 2.4$. The area from $z = 0$ to $z = .2$ is $.0793$, and the area from $z = 0$ to $z = 2.4$ is $.4918$. So to find the area between $z = .2$ and $z = 2.4$, we take the difference between the two areas (not between the two z's) and we find there is $.4918 - .0793 = .4125 = 41.25$ per cent of the sales between \$14,000 and \$25,000. Graphically, we have determined area I and area II and subtracted area I from area II to find the desired area (Figure 6.2.7). Instead of subtracting the areas, could we have just subtracted $.2$ from 2.4 and looked up the area for $z = 2.2$? If we do this, we get $.4861$—which is

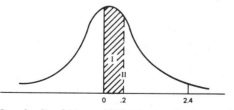

Figure 6.2.7. *Standardized Normal Distribution Region to the Right of the Mean*

not the answer we found. Do you see that the areas are different? Compare the two areas in Figure 6.2.8 to be certain that they are not the same.

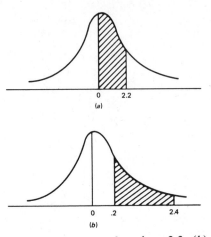

Figure 6.2.8. (*a*) *Area Between z = 0 and z = 2.2;* (*b*) *Area Between z = 0.2 and z = 2.4*

The four types of problems we just examined can be generalized into the following categories and approaches to determining proportions of areas under the curve:

1. Area between $z = 0$ and z_0 (z_0 is to the right of $z = 0$) is found by directly using Table III. If the z_0-value falls to the left of $z = 0$, that is, it is negative, use the symmetric property of the normal distribution and look up the proportion for the positive z_0-value (Figure 6.2.9).

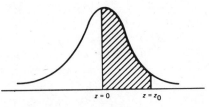

Figure 6.2.9. *Area Between z = 0 and z_0*

2. If z_0 is greater than 0, and the area from that z_0 and above is desired, since 50 per cent of the area lies in each half, find the area from $z = 0$ to the specified z_0 and then subtract this value from .5000. For example, if $z_0 = 2.3$, then the proportion of scores above z_0 is $.5000 - .4893 = .1107$ (Figure 6.2.10).

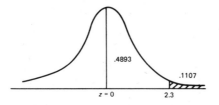

Figure 6.2.10. *Area Between z = 0 and z = 2.3*

3. If the area between two z-values is required, and one z-value falls to the left of $z = 0$, and the other z-value falls to the right, the total area is found by adding the separate areas. For example, the area between $z = -1$ and $z = 1.5$ is found by finding the area from $z = -1$ to $z = 0$ (the same as $z = 0$ to $z = 1$), which is .3413, and the area from $z = 0$ to $z = 1.5$ is .4332, so the area from $z = -1$ to $z = 1.5$ is $.3413 + .4332 = .7745$ (Figure 6.2.11).

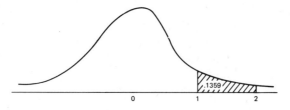

Figure 6.2.11. *Area Between z = -1 and z = 1.5*

4. If both z-values are to the left or to the right of $z = 0$, the area between the z-values is found by taking the difference between the larger area and the smaller area (*not* the larger z-value and the smaller z-value). For example, the area between $z = 1$ and $z = 2$ is found by taking the difference between .4772 and .3413, that is, .1359 (Figure 6.2.12).

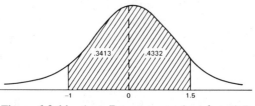

Figure 6.2.12. *Area Between z = 1 and z = 2*

The area under a normal distribution curve can also be interpreted in terms of probability measures. That is, if we know or can assume that a collection of scores are normally distributed, then we can speak of the chance of some event occurring as well as the per cent or proportion of the distribution satisfying a particular outcome.

EXAMPLE 1

We can reexamine the sales of the soft-drink company that were normally distributed with $\mu = 13,000$ and $\sigma = 5,000$, and respond to the following questions:

(a) What is the chance that a week chosen randomly will have sales of at least $13,000 and at most $20,000?

(b) At least $27,000?

(c) At least $9,000 and at most $15,000?

(d) At least $20,000 and at most $25,000?

The solutions can be found by following the general approach just considered.

Solutions

(a) $P(13,000 \leq x \leq 20,000) = ?$ Since $P(x = 13,000) = P(z = 0)$ and $P(x = 20,000) = P(z = 1.4)$, we want to find $P(0 \leq z \leq 1.4) = P(13,000 \leq x \leq 20,000)$ after the distribution has been standardized. We find $P(0 \leq z \leq 1.4) = .4192$. That is, there is almost a 42 per cent chance that the sales in a randomly chosen week would be between $13,000 and $20,000.

(b) $P(x \geq 27,000) = P(z \geq 2.8)$, since $z = \dfrac{27,000 - 13,000}{5,000} = \dfrac{14,000}{5,000} =$ 2.8. Therefore, $P(x \geq 27,000) = .5000 - .4974 = .0026$.

(c) $P(9,000 \leq x \leq 15,000) = P(-.8 \leq z \leq .4) = .2881 + .1554 = .4435$.

(d) $P(20,000 \leq x \leq 25,000) = P(1.4 \leq z \leq 2.4) = P(z \leq 2.4) - P(z \leq 1.4)$ $= .4918 - .4192 = .0726$.

PROBLEMS

6.2.1 Over a period of 6 months, 500 customers were timed how long they had to wait for a Super Burger. It was found that on the average the wait was 180 seconds, with a standard deviation of 30 seconds, and the waiting distribution was normally distributed about the mean.

(a) What is the probability that someone had to wait at most 4 minutes (4 minutes or less)?

(b) What is the proportion of people that had to wait at most 3 minutes?

(c) What is the probability that someone waited for more than $4\frac{1}{2}$ minutes?

(d) How many people waited between 2 and 4 minutes?

(e) What percentage of people waited between $3\frac{1}{2}$ and 4 minutes?

(f) What do you think will be the response of a Super Burger salesperson if you say you have waited more than 3σ from the mean and, as this occurs only .0013 of the time, wonder whether this is due to chance or to a machine and/or person failure?

(g) How many of the 500 people had to wait for more than 2 minutes?

(h) What x-value is represented by a $z = 1.5$? What percentile does this x-value represent; that is, what per cent of the distribution is less than this x-value?

6.2.2 The average seasonal rainfall in a certain town is 18.8 inches with a standard deviation of 6.5 inches. Assuming that the seasonal rainfall for this town is normally distributed, in how many out of 50 years would we expect between 15.0 and 25.0 inches of rain to fall?

6.2.3 The lifetimes of certain kinds of electronic devices have a mean of 270 hours and a standard deviation of 26 hours. Assuming that the distribution of these lifetimes, which are measured to the nearest hour, can be approximately that of a normal distribution, find:

(a) The probability that any one of these devices will have a lifetime of from 260 to 280 hours.

(b) The percentage that will have lifetimes of at least 300 hours.

(c) The value below which we will find the lowest 8 per cent of lifetimes.

6.2.4 Sketch the normal distribution and shade in the appropriate regions discussed in the soft-drink company problem.

6.3 Normal Approximation to the Binomial Distribution

In this section we shall examine how and when problems involving the binomial distribution can be solved quite accurately by using the normal distribution as an approximation. Recall that in our examination of the binomial distribution, Table II contained probability values for sample sizes up to $n = 25$. These tables are insufficient for determining probabilities for $n > 25$. As it turns out, the standard normal distribution provides a close approximation to the binomial distribution for $n > 25$—the larger the value of n, the better the approximation.

Consider the situation where an employment agency states that 40 per cent of their clients get hired. We shall assume that the chance that one individual gets hired is independent of another (under what circumstances can this assumption be made?) and determine the binomial probability distribution for a sample of $n = 10$ clients. The number of clients obtaining a job and the associated probabilities are listed in Table 6.3.1.

If we use the possible number of hired as the class marks, the histogram would appear as shown in Figure 6.3.1.

In order to determine a normal distribution as an approximation to this particular binomial distribution, it seems reasonable to use the μ and σ of the binomial distribution as the μ and σ of the normal distribution. We

Table 6.3.1. Binomial Probability
for $n = 10$ and $p = .40$

Number hired	P(number hired)
0	.006
1	.040
2	.121
3	.215
4	.251
5	.201
6	.111
7	.042
8	.011
9	.002
10	.000
	1.000

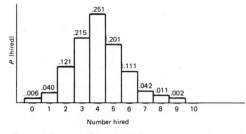

Figure 6.3.1. *Histogram for $n = 10$ and $p = .40$*

have then that $\mu = np = 10(.40) = 4$ and $\sigma = \sqrt{npq} = \sqrt{10(.4)(.6)} = 1.55$. Since the normal distribution is applicable for continuous variables, while the binomial distribution is used for discrete variables, before we determine the closeness of this normal to this binomial, we agree to treat a discrete score such as number hired $= 6$, to lie in the continuous interval 5.5 to 6.5. That is, $P(x = 6) = P(5.5 \le x \le 6.5)$. Now for the comparison:

Binomial		Normal	
$p = .4;$	$n = 10$	$\mu = 4;$	$\sigma = 1.55$

1. What is the chance that at least 6 of the 10 will be hired?
Binomial: $P(X \ge 6) = .111 + .042 + .011 + .002 = .166$.
Normal: To determine the normal approximation, we want to find the area under the curve from $X = 5.5$ to the right (Figure 6.3.2). The x-score must be converted to a z-score: $z = \dfrac{5.5 - 4}{1.55} = \dfrac{1.5}{1.55} = .97$; that is, $P(X \ge$

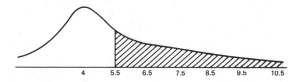

Figure 6.3.2. *Area Under the Curve from X = 5.5 to the Right*

$5.5) = P(Z \geq .97)$. Since the area we want to find is from a value to the right of the mean, we subtract the area found in Table III where $z = .97$, namely .3340 from .5000 (the area to the right of the mean), and find $.5000 - .3340 = .1660$. This value is exactly the same, to three places, as we found using the binomial distribution.

2. Find the chance that between 2 and 6 of the next 10 people sent out for jobs are hired.

Binomial: $P(2 \leq x \leq 6) = .121 + .215 + .251 + .201 + .111 = .899$.

Normal: We must find the area under the curve between $x = 1.5$ and $x = 6.5$, that is, $P(1.5 \leq x \leq 6.5)$. First, we convert to z-scores: $x = 1.5$ becomes $z_1 = \dfrac{1.5 - 4}{1.55} = \dfrac{-2.5}{1.55} = -1.61$; $x = 6.5$ becomes $z_2 = \dfrac{6.5 - 4}{1.55} = \dfrac{2.5}{1.55} = 1.61$. Thus $P(1.5 \leq x \leq 6.5) = P(-1.61 \leq z \leq 1.61)$. We shall determine the shaded area in Figure 6.3.3. Since the curve is symmetric, we find

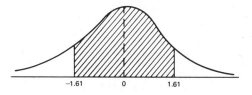

Figure 6.3.3. *Area Between −1.61 and 1.61*

the area from $z = 0$ to $z = 1.61$ and double it to obtain the total area. The area between $z = 0$ and $z = 1.61$ is .4463; thus the normal approximation to the exact binomial value is $2(.4463) = .8926$, which is a pretty good approximation to the binomial value of .899.

The reason that the two approximations we made were so accurate, although n was small, was because p was close to .50. The closer p is to .50, the closer the binomial distribution appears to approximate normal (e.g., see the two binomial distributions in Table 4.6.1). Some authors use the rule that the normal distribution to the binomial exact value will be good if (after choosing the smaller probability of p or q, say it is p) $np > 5$, while others use the more conservative estimate that $np > 10$.

PROBLEMS

6.3.1 A survey found that gas consumption for one of the high-powered sports cars is normally distributed with a mean of 9.2 mpg and a standard deviation of 2.4 mpg. If one of these cars is purchased, find the probability that:
  (a) It will get at least 9.2 mpg.
  (b) It will get at least 10.6 mpg.
  (c) It will get at most 7.5 mpg.

6.3.2 If the heights of 300 students are normally distributed with $\mu = 68$ inches and $\sigma = 3$ inches, determine:
  (a) The standardized score of a student whose height is 72 inches.
  (b) How many students in this group have heights of at least 68 inches?
  (c) How many students in this group have heights of at most 64 inches?
  (d) How many students in this group have heights between 65 and 71 inches?
  (e) What is the chance that a student chosen at random from this group will be at least 76 inches?
  (f) What is the probability that a student chosen at random from this group will have a height that is within 1 standard deviation of the mean?

6.3.3 If 40 per cent of car sales consists of used cars, find the exact binomial probability and the normal approximation to the binomial of the next 20 sales:
  (a) At most 6 will be used cars.
  (b) At least 10 will be used cars.
  (c) Between 5 and 9 new cars will be sold.

6.3.4 The mean grade on a final exam was 72 and the standard deviation was 9. If the top 10 per cent are to receive grades of A, what is the minimum grade a student must score in order to receive an A?

6.4 Hypothesis Testing

In testing hypotheses made about normal populations, the form of the test is the same as we used in testing hypotheses made about populations that were binomially distributed:
  1. Create the null and alternative hypotheses; H_0 and H_A.
  2. State the level of significance and the sample size.

3. Determine a decision rule based upon the alternative hypothesis and the level of significance. The form of the decision rule is: Reject the null hypothesis if the number of successes falls in the critical region.
4. Test the null hypothesis with the sample data.
5. State that the null hypothesis is rejected or not rejected depending upon the sample data and the critical region.

The critical and acceptance regions are determined by the chosen level of significance. For example, if a two-tailed hypothesis were being tested and $\alpha \leq .05$, we would want the critical region to be divided so that .025 was in each tail (Figure 6.4.1). From Table III, we see that if $z = 1.96$, that

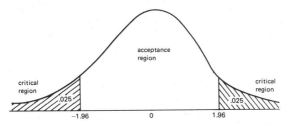

Figure 6.4.1. *Critical and Acceptance Regions for a Two-Tailed Test and $\alpha \leq .05$*

is, 1.96 standard deviations from the mean, $z = 0$, that .4750 of the area lies between $z = 0$ and $z = 1.96$. Thus $P(z \geq 1.96) = .5000 - .4750 = .0250$. We obtain, then, that a two-tailed hypothesis will be rejected if the sample data yield a z-score either at least 1.96 or at most -1.96, since z-scores greater than 1.96 or less than -1.96 will occur at most $.025 + .025 = .05$ of the time. If the value of z is such that $-1.96 < z < 1.96$, the hypothesis cannot be rejected, since the value falls in the acceptance region. For example, a $z = 1.43$ would not be rejected for a two-tailed test if $\alpha \leq .05$, since it would fall in the interval $(-1.96, 1.96)$.

If the hypothesis were one-tailed to the right, the critical region would be determined by that z-value above which 5 per cent of the distribution occurs (Figure 6.4.2). From Table III, we see that if $z = 1.64$, then .4495 of the area lies between $z = 0$ and $z = 1.64$. Thus $P(z \geq 1.64) = .5000 - .4495 = .0505$. We then obtain that the one-tailed hypothesis to

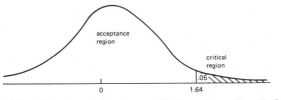

Figure 6.4.2. *Critical and Acceptance Regions for a One-Tailed Test and $\alpha \leq .05$*

the right will be rejected if the sample data yield a z-score that is at least 1.64. If the z-score is less than 1.64, the hypothesis cannot be rejected.

For one-tailed tests to the left, since the distribution is symmetric, a z-value less than -1.64 will determine that the hypothesis should be rejected. That is, since $P(z \geq 1.64) = P(z \leq -1.64) = .05$, any z-value less than -1.64 will occur at most 5 per cent of the time. You have probably noticed that at $z = 1.64$ the area determined is not exactly 5 per cent of the area; rather, it is .0505. While some authors use 1.64 and others 1.65, we will use the value of 1.64 when $\alpha = .05$ and a one-tailed test is required. What reason might someone use to argue for 1.64? For 1.65?

Before we can begin testing hypotheses regarding normal populations, it is necessary first to discuss a sampling distribution as versus a population distribution. A population distribution has a unique mean and standard deviation. That is, if we took as the population all the baseball players on the 1956 New York Yankees, the mean batting average for the entire team would be the population mean. However, if we took a random sample from this population, the sample would determine a particular sample mean, $\bar{x}$, and standard deviation, s, which would most probably not be the same as the mean and standard deviation of the population distribution, μ and σ. Further, if we took another sample, after replacing the first collection of sample data, the second sample would determine a second sample mean and standard deviation that was probably not the same as either the first sample mean and standard deviation or the population mean and standard deviation. Inasmuch as we will be making judgments about the population mean based on a particular sample mean, we need to know what the mean and standard deviation would be of all the possible samples of size n that we could take. In other words, when we convert from x-scores to z-scores, we use the conversion formula $z = (x - \mu_x)/\sigma_x$, where μ_x and σ_x are determined by the population distribution of x-scores. But, when we will be converting based on a sample, we need to know the mean and standard deviation of the population of all samples that could be taken; that is, we need to determine $\mu_{\bar{x}}$, the mean of all the possible means, and $\sigma_{\bar{x}}$, the standard deviation of all the possible means. If we knew the population mean and standard deviation, there would be no need to take any random samples in order to gain insight into the population. Since, in most cases, it is too time consuming and/or costly to determine the population mean and standard deviation, we take random samples from the population and make inferences about the population based upon the sample findings.

The following will demonstrate the remarkable relationship between the population mean μ_x and the sampling distribution mean $\mu_{\bar{x}}$, and between the population standard deviation σ_x and the sampling distribution standard deviation $\sigma_{\bar{x}}$. Suppose the ages of three children are 3, 4, and 5, and that these scores constitute the population. We find that

$$\mu_x = \frac{3+4+5}{3} = 4, \quad \text{and} \quad \sigma_x = \sqrt{\frac{(3-4)^2+(4-4)^2+(5-4)^2}{3}} = .8165$$

Now, if we take all possible samples of size $n = 1$, replacing each one after each pick, we have three samples: 3, 4, and 5—that is, each element in the population. Obviously, the means of each of these samples are also 3, 4, and 5, such that $\mu_{\bar{x}} = 4$ and $\sigma_{\bar{x}} = .8165$.

But before we jump to any conclusions such as $\mu_x = \mu_{\bar{x}}$ and $\sigma_x = \sigma_{\bar{x}}$, consider the samples when $n = 2$ and when $n = 3$ (Tables 6.4.1 and 6.4.2).

Table 6.4.1. Sample Distribution for $n = 2$

Sample	$\bar{x}$
3, 3	3
3, 4	3.5
3, 5	4
4, 3	3.5
4, 4	4
4, 5	4.5
5, 3	4
5, 4	4.5
5, 5	5

$$\mu_{\bar{x}} = \frac{36}{9} = 4$$

$$\sigma_{\bar{x}} = .5773$$

Table 6.4.2. Sample Distribution for $n = 3$

Sample	$\bar{x}$	Sample	$\bar{x}$	Sample	$\bar{x}$
3, 3, 3	$\frac{9}{3}$	4, 4, 4	$\frac{12}{3}$	5, 5, 5	$\frac{15}{3}$
3, 4, 3	$\frac{10}{3}$	4, 3, 4	$\frac{11}{3}$	5, 3, 5	$\frac{13}{3}$
3, 3, 4	$\frac{10}{3}$	4, 4, 3	$\frac{11}{3}$	5, 5, 3	$\frac{13}{3}$
3, 5, 3	$\frac{11}{3}$	4, 5, 4	$\frac{13}{3}$	5, 4, 5	$\frac{14}{3}$
3, 3, 5	$\frac{11}{3}$	4, 4, 5	$\frac{13}{3}$	5, 5, 4	$\frac{14}{3}$
3, 4, 4	$\frac{11}{3}$	4, 3, 3	$\frac{10}{3}$	5, 3, 3	$\frac{11}{3}$
3, 5, 5	$\frac{13}{3}$	4, 5, 5	$\frac{14}{3}$	5, 4, 4	$\frac{13}{3}$
3, 4, 5	$\frac{12}{3}$	4, 3, 5	$\frac{12}{3}$	5, 3, 4	$\frac{12}{3}$
3, 5, 4	$\frac{12}{3}$	4, 5, 3	$\frac{12}{3}$	5, 4, 3	$\frac{12}{3}$

$$\mu_{\bar{x}} = \frac{324/3}{27} = 4 \qquad \sigma_{\bar{x}} = .4714$$

In each of the three samples we find that $\mu_x = \mu_{\bar{x}}$; that is, the population mean equals the mean of all the sample means for any n. This is, in fact, true in general, but will not be proved here. Although it is not apparent, there is a relationship between σ_x and $\sigma_{\bar{x}}$. It can be proved that, in general,

$\dfrac{\sigma_x}{\sqrt{n}} = \sigma_{\bar{x}}$. In our three samples we had when $n = 1$, $\sigma_{\bar{x}} = .8165$, such that

$\dfrac{.8165}{\sqrt{1}} = .8165$; when $n = 2$, we find that $\dfrac{.8165}{\sqrt{2}} = .5773$; and when $n = 3$, we

find that $\dfrac{.8165}{\sqrt{3}} = .4714$. If we had not taken all the possible samples for

each n, the relationships would be approximately true and would get better as the number of samples increases. These interesting relationships allow

us to use the conversion from x- to z-scores, where $z = \dfrac{\bar{x} - \mu_{\bar{x}}}{\sigma_{\bar{x}}} = \dfrac{\bar{x} - \mu_x}{\sigma_x / \sqrt{n}}$.

Before we consider some examples that will demonstrate applications of our findings, there is another exciting implication suggested by the sampling distributions we obtained. Notice that the population scores determine a rectangular distribution, that is, the scores were 3, 4, and 5, and each had a frequency of 1. What about the sample distributions—what shapes do they take? For $n = 1$, we again have the rectangular distribution. For $n = 2$, we have scores of 3, 3.5, 4, 4.5, and 5 with frequencies of 1, 2, 3, 2, 1—what type of distribution is suggested by these frequencies? Create a frequency distribution for the sample means when $n = 3$ and determine the shape of the frequency distribution. You should find them both to be normal distribution shapes. That is, given a rectangular distribution of population scores, the sampling distribution of all sample of size n will approach a normal distribution. The approximation gets better as n increases. As a matter of fact, the initial distribution does not have to be rectangular. This amazing relationship is known as the *Central Limit Theorem*. The theorem states: A collection of random sample distributions will approach a normal distribution shape as n, the sample size, gets large. The expected value of the sampling distribution of means will be μ, and the

standard deviation of the means will be $\dfrac{\sigma}{\sqrt{n}}$, where μ is the mean of the

original population distribution and σ is the standard deviation of the population distribution. This relationship allows us to use the normal curve as an approximation of a sampling distribution of $\bar{x}$. The approximation gets better as n increases. The following examples should help clarify our findings.

EXAMPLE 1

If a random sample of size $n = 100$ is drawn from a population having $\mu = 9$ and $\sigma = 5$, we can determine probabilities of outcomes that the

mean, $\bar{x}$, satisfies certain conditions before looking at our sample data. For example, we will find the probabilities that the sample mean will be:
(a) At most 10.
(b) At least 8.5.
(c) Lie between 8 and 9.5.

Since $\mu_x = 9$ and n is large, we can say that $\mu_x = \mu_{\bar{x}} = 9$, and too, $\sigma_{\bar{x}} = \dfrac{\sigma_x}{\sqrt{n}} =$

$\dfrac{5}{\sqrt{100}} = .5$. The sampling distribution of $\bar{x}$ can be approximated by the

normal distribution, since n is 100. We have then that $z = \dfrac{\bar{x} - \mu_x}{\sigma_x / \sqrt{n}} = \dfrac{\bar{x} - 9}{.5}$.

In other words, we have that $\bar{x} = .5z + 9$.

Solution
(a) $P(\bar{x} \le 10) = P(.5z + 9 \le 10) = P(z \le 2) = .9773$.
(b) $P(\bar{x} \ge 8.5) = P(.5z + 9 \ge 8.5) = P(z \ge -1) = .8413$.
(c) $P(8 \le \bar{x} \le 9.5) = P(8 \le .5z + 9 \le 9.5) = P(-1 \le .5z \le .5)$
    $= P(-2 \le z \le 1) = P(z \le 1) - P(z \le -2) = .8413 - .0228 = .8185$.

We shall use the form of the hypothesis test we have been using and the relationships just discussed in examining the following situations.

EXAMPLE 2

A weight-reducing program states that in 2 weeks' time an average person will lose on the average 10 pounds if the program is followed exactly. Thirty people who followed the program exactly have lost on the average 8.2 pounds, with a standard deviation of 3.5 pounds. The question is whether the stated population mean, $\mu = 10$, differs significantly from the sample mean, $\bar{x} = 8.2$. That is, can we claim that the sample findings are reasonable to expect, or are the sample findings different enough from the population findings to cast doubt on the claim?

The test analysis is:
1. $H_0 : \mu = 10$.
   $H_A : \mu \neq 10$; that is, $\mu < 10$ or $\mu > 10$. (Although the sample mean was 8.2, the null and alternative hypotheses are set up *before* the data are collected.)
2. $\alpha \le .05$ and $n = 30$.
3. Our decision rule is: Reject the null hypothesis, $\mu = 10$, if the z-value determined by the sample mean of 8.2 and the sample standard deviation of 3.5 falls in the critical region. Since the alternative hypothesis is two-tailed, we want to find the z-scores associated with the critical regions, at the extremes of the normal distribution, which contain at most 2.5 per cent ($\frac{1}{2}$ or 5 per cent) of the total area. From our earlier findings, we know that if we find $z > 1.96$ or $z < -1.96$, then the value has fallen in the critical region, which contains at most

5 per cent of the area (see Figure 6.4.1). The decision rule is: Reject
the null hypothesis, $\mu = 10$, if the z-score, $z = \dfrac{\bar{x} - \mu_{\bar{x}}}{\sigma_{\bar{x}}}$, falls in the
critical region, that is, if this z-score is greater than 1.96 or less than
−1.96.
4. The null hypothesis test is found by substituting 8.2 for $\bar{x}$ and 10 for
   $\mu_x = \mu_{\bar{x}}$. However, the sample standard deviation, $s = 3.5$, is not
   equivalent to σ_x. But, when n is at least 25, we can use s, the sample
   standard deviation as an approximation for σ_x. We have then that

$$z = \frac{\bar{x} - \mu_{\bar{x}}}{\sigma_{\bar{x}}} = \frac{\bar{x} - \mu}{\sigma_x/\sqrt{n}} \approx \frac{\bar{x} - \mu}{s/\sqrt{n}} = \frac{8.2 - 10}{3.5/\sqrt{30}} = \frac{-1.8}{.64} = -2.81$$

This value of $z = -2.81$ fall in the left critical region (Figure 6.4.3).

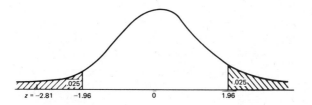

z = -2.81   -1.96   0   1.96

Figure 6.4.3. *Critical Region for $\alpha \le .05$ Two-Tailed Test*

5. The null hypothesis, $\mu = 10$, is rejected as the obtained z-value,
   $z = -2.81$, falls to the left of $z = -1.96$. The chance that a z-value
   would fall in the critical region if the null hypothesis was true is at
   most 5 per cent. Thus we are led to reject the statement that the
   population average loss is 10 pounds.

P R O B L E M

6.4.1 Would this null hypothesis have been rejected if the alternative
hypothesis had been $\mu < 10$? (Why?)

EXAMPLE 3
A study was done to determine if an antismoking campaign had per-
suaded people to stop or at least cut down on smoking. Prior to the
campaign, a national survey found that a smoker smoked an average of 11
cigarettes per day. After the campaign, a sample of 100 smokers resulted in
the findings that the average number of cigarettes smoked per day as 10,
with a standard deviation of 3. A cigarette company wonders if the
difference in means of 1 cigarette is significant, or if it is due to chance.

They decide to set up and test the following hypothesis:
1. $H_0 : \mu = 11$; $H_A : \mu < 11$.
2. $\alpha \le .05$; $n = 100$.
3. Reject the null hypothesis, $\mu = 11$, if the z-value determined by $\bar{x} = 10$ and $s = 3$ falls in the critical region, that is, if the calculated z-value is less than -1.64.
4. $z = \dfrac{\bar{x} - \mu_{\bar{x}}}{\sigma_{\bar{x}}} = \dfrac{10 - 11}{\sigma_x / \sqrt{n}} = \dfrac{-1}{3/\sqrt{100}} = \dfrac{-1}{3/10} = -\dfrac{10}{3}$.
5. Since the sample $z = -3.33$ falls in the critical region to the left of $z = -1.64$, the null hypothesis is rejected and we can be 95 per cent confident $(1 - \alpha = \text{degree of confidence})$ that the findings demonstrate a significant change in smoking habits.

PROBLEM

6.4.2 Why wasn't the alternative hypothesis set at $\mu > 11$ instead of $\mu < 11$?

EXAMPLE 4

Some people believe that gasoline shortages and higher prices for gasoline have changed the driving habits of many people. A survey was taken to test if people drive slower after a price rise in gasoline. Before the shortage, a national survey found that highway driving speeds averaged 58 mph where the speed limit was 60 mph. After the shortage, the average speed of 50 drivers was found to be 56 mph, with a standard deviation of 9 mph. Can it be stated that driving habits have changed as a result of the increase in cost of gasoline? The following test will examine this situation.
1. $H_0 : \mu = 58$ vs. $H_A : \mu < 58$.
2. $\alpha \le .05$; $n = 50$.
3. Reject the null hypothesis, $H_0 : \mu = 58$, if the z-value determined by $\bar{x} = 56$ and $s = 9$ is less than $z = -1.64$.
4. $z = \dfrac{\bar{x} - \mu_{\bar{x}}}{\sigma_{\bar{x}}} = \dfrac{\bar{x} - \mu}{\sigma_x} \approx \dfrac{56 - 58}{9/\sqrt{50}} = \dfrac{-2}{9/7.1} = \dfrac{-2(7.1)}{9} = \dfrac{-14.2}{9} = -1.58$.
5. The hypothesis cannot be rejected, since $-1.58 > -1.64$. Thus, he will accept the null hypothesis that the increase in gasoline costs has not changed driving habits.

The three examples we have just examined have the same rationale. That is, we have shown that if a population has a mean μ_x and a standard deviation σ_x, then the sampling distribution of $\bar{x}$, which is determined by taking all possible samples of size n, will have a mean, $\mu_{\bar{x}} = \mu_x$ and a standard deviation, $\sigma_{\bar{x}} = \dfrac{\sigma_x}{\sqrt{n}}$. Suppose, for example, we are given that a population distribution has $\mu_x = 100$ and $\sigma_x = 10$. We know that if we took all possible samples of $n = 49$ from this population, the sampling distribution mean, $\mu_{\bar{x}} = \mu_x = 100$—that is, the mean of all the means determined by samples of size $n = 49$ would be the same as the population mean, $\mu_x = 100$; and the standard deviation of all the means of the sampling distribution would be $\sigma_{\bar{x}} = \dfrac{\sigma_x}{\sqrt{n}} = \dfrac{10}{\sqrt{49}} = \dfrac{10}{7} = 1.43$. If our null hypothesis was $\mu_x = 100$, and this was correct, the sampling distribution would be approximately normal with a mean of 100 and a standard deviation of 1.43 (Figure 6.4.4).

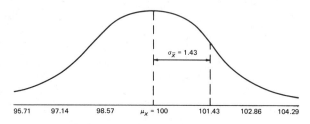

Figure 6.4.4. *Sampling Distribution*

When we take a single sample of size $n = 49$, we could find any mean, $\bar{x}$. Some means are more probable than others. For example, an $\bar{x} = 99$ is much more probable than an $\bar{x} = 96$; that is, the $\bar{x} = 99$ is closer to the sampling distribution mean, $\mu_x = 100$ than is $\bar{x} = 96$. If we took a sample of size $n = 49$ and obtained $\bar{x} = 99$, we would not reject the hypothesis. But, if we found that $\bar{x} = 96$, since it is more than 2 standard deviations from the mean, we might very well reject the null hypothesis that $\mu_x = 100$. In the gasoline problem, we had $\mu_x = 11$, $\bar{x} = 10$, and $\sigma_{\bar{x}} = .30$. If $\mu_x = 11$, then so would $\mu_{\bar{x}} = 11$; and then $\mu_{\bar{x}} - 3\sigma_{\bar{x}} = 11 - 3(.30) = 10.10$. That is, 3 standard deviations to the left of $\mu_{\bar{x}}$ is 10.10. Our sample mean was $\bar{x} = 10$. This value would fall to the left of 10.10. It would be more than 3 standard deviations from the population mean $\mu_x = 11$. Since it is so far away, we reject the null hypothesis that $\mu_x = 11$ as an $\bar{x} = 10$ is very improbable when $\mu_{\bar{x}} = 11$ and $\sigma_{\bar{x}} = .30$.

EXAMPLE 5

(a) Suppose a claim is made that a normal population has $\mu_x = 100$, $\sigma_x = 10$, and a random sample of $n = 64$ determines $\bar{x} = 105$. Should we reject or retain the hypothesis with a one-tailed test to the right?

1. $H_0 : \mu = 100$; $H_A : \mu > 100$.
2. $\alpha \le .05$; $n = 64$.
3. Reject the null hypothesis if the determined z is greater than 1.64.
4. $z = \dfrac{\bar{x} - \mu_{\bar{x}}}{\sigma_{\bar{x}}} = \dfrac{\bar{x} - \mu_x}{\sigma_{\bar{x}}/\sqrt{n}} = \dfrac{105 - 100}{10/\sqrt{64}} = \dfrac{5}{10/8} = \dfrac{40}{8} = 4.$

5. Reject the hypothesis since $4 > 1.64$, and this value falls in the .05 critical region in the right tail. Note that in this case σ_x was known and we did not need to approximate σ_x by s.

(b) Suppose again that $\mu_x = 100$, $\sigma_x = 10$, but this time a sample of $n = 9$ results in $\bar{x} = 105$. If $z > 1.64$, the null hypothesis is rejected. But $z = \dfrac{105-100}{10/\sqrt{9}} = \dfrac{5}{10/3} = 1.5$; and since $1.5 < 1.64$, the hypothesis cannot be rejected. We can use the normal distribution tables, although n is 9 (small), since we know the population σ_x and do not have to approximate it with s. In Section 6.6 we shall consider an alternative approach when our sample size is small and the population standard deviation σ_x is not known.

PROBLEMS

6.4.3 Why do you think that in Example 5(b) the hypothesis cannot be rejected but in Example 5(a) it is? How do you explain the effects of the sample size?

6.4.4 Test the hypothesis that $\mu = 50$ against $H_A : \mu < 50$ or $\mu > 50$, if a sample of $n = 36$ produces a sample mean of 52 and standard deviation, $s = 10$, with $\alpha \leq .05$.

6.4.5 Test the hypothesis that $\mu = 200$ against $H_A : \mu > 200$, if a sample of $n = 100$ yielded a sample $\bar{x} = 210$ and $s = 25$, and $\alpha \leq .05$; is it necessary to test again for $\alpha \leq .10$? (Why?)

6.4.6 Test the hypothesis that $\mu_x = 100$, where $\sigma_x = 15$, $n = 25$, and $\alpha \leq .05$ with a two-tailed test if:
(a) $\bar{x} = 92$.
(b) $\bar{x} = 108$.
(c) $\bar{x} = 96$.

6.4.7 A random sample of size 100 is to be drawn from a population having $\mu = 8$ and $\sigma = 5$. What is the probability, prior to taking the sample, that $\bar{x}$ will be:
(a) At most 9?
(b) At least 7.5?
(c) Between 7.8 and 8.2?

6.5 Estimation

In Section 5.3 we discussed making estimates about population proportions based upon sample proportions. The *point estimate* of the population proportion, p, is the sample proportion, $\bar{p}$. An *interval estimate* of the population proportion is an interval that we can construct so as to have some degree of confidence that the interval contains the population proportion. For example, a 95 per cent confidence interval for the population proportion was seen to be $(\bar{p} - 1.96\sigma_{\bar{p}}, \bar{p} + 1.96\sigma_{\bar{p}}) = \left(\bar{p} - 1.96\dfrac{\sigma_p}{\sqrt{n}}, \bar{p} + 1.96\dfrac{\sigma_p}{\sqrt{n}}\right)$. In this section we shall discuss confidence intervals for population means. The approach is the same as in our earlier discussion of confidence intervals for population proportions. For example, the 95 per cent confidence interval of a population mean is $\left(\bar{x} - 1.96\dfrac{\sigma_x}{\sqrt{n}}, \bar{x} + 1.96\dfrac{\sigma_x}{\sqrt{n}}\right)$, while the 90 per cent confidence interval for the population mean is $\left(\bar{x} - 1.64\dfrac{\sigma_x}{\sqrt{n}}, \bar{x} + 1.64\dfrac{\sigma_x}{\sqrt{n}}\right)$. The change in the degree of confidence is dependent on the change in the number of standard deviations from the mean. We can be 95 per cent confident with the first interval, since we have chosen 1.96 standard deviations from the sample mean, while we can be 90 per cent confident with the second interval, which extends only 1.64 standard deviations from the sample mean. In other words, 95 per cent of the area of a normal distribution is within 1.96 standard deviations from the mean, while 90 per cent of the area of a normal distribution is within 1.64 standard deviations from the mean.

The following situations will help to clarify the usage and value of confidence-interval estimates of population means, based upon sample findings.

EXAMPLE 1

1. A supermarket manager wants to determine if it would be profitable to stay open all night. He decides that rather than ask a random sample of customers if they would shop late at night, he will stay open five nights a week for 6 weeks. He finds that over the $n = 30$ days the distribution of people was approximately normally distributed with the average number of customers $= 56$ per night, with a standard deviation of 8 customers per night. He wants to determine a 90 per cent confidence interval for the population mean.

Since a 90 per cent interval is chosen, 5 per cent of the distribution will fall outside to the left of the interval and 5 per cent will fall to the right, as he is willing to allow an $\alpha \le .10$. The corresponding z-value is 1.64. Although he does not know the population standard deviation, n was large enough so as to approximate the population standard deviation with the sample standard deviation. His sample data yielded $\bar{x} = 56$ and $s = 8$. We have then that

$$\left(\bar{x} - 1.64\frac{\sigma_x}{\sqrt{n}}, \bar{x} + 1.64\frac{\sigma_x}{\sqrt{n}}\right) = \left(56 - \frac{1.64(8)}{\sqrt{30}}, 56 + \frac{1.64(8)}{\sqrt{30}}\right)$$

Using the square-root table and simplifying, we arrive at $(56 - 1.64(1.46), 56 + 1.64(1.46))$, which becomes $(56 - 2.39, 56 + 2.39)$. As such, the 90 per cent confidence interval about the population mean is $(53.61, 58.39)$. There is a 90 per cent chance that this interval contains the true population mean.

The 95 per cent confidence interval for this situation would be found to be

$$\left(56 - \frac{1.96(8)}{\sqrt{30}}, \ 56 + \frac{1.96(8)}{\sqrt{30}}\right)$$

$$= (56 - 1.96(1.46), \ 56 + 1.96(1.46))$$

$$= (56 = 2.86, \ 56 + 2.86)$$

$$= (53.14, 58.86)$$

Although there is an increase in degree of confidence for this interval over the first one we found, the width of the interval is also larger. So, while we have more confidence, we also have a wider, less exact, interval. The easiest interval to create is the 100 per cent confidence interval: $(0, \infty)$. In this situation, even though we are 100 per cent confident that this interval contains the population mean, it is, in fact, valueless.

P R O B L E M S

6.5.1 (a) What would be the 99 per cent confidence interval for the supermarket problem?

(b) Locate the 90, 95, and 99 per cent confidence intervals on a sketch of a normal distribution curve.

6.5.2 If the doctor told Mario and Jackie that the baby would be born on Christmas Day, what type of estimate was given?

EXAMPLE 2

The weight-reducing program tried by 30 people determined an average loss of 8.2 pounds, with a standard deviation of 3.5 pounds. The 99 per cent confidence interval can be found by using the fact that 99 per cent of the area is within 2.58 standard deviations of the mean in a standardized normal distribution. We find then that the 99 per cent confidence interval is

$$\left(\bar{x} - 2.58\frac{\sigma_x}{\sqrt{n}}, \ \bar{x} + 2.58\frac{\sigma_x}{\sqrt{n}}\right)$$

As $\bar{x} = 8.2$, $\sigma_x \approx s = 3.5$, and $n = 30$, we have that

$$\left(8.2 - \frac{2.58(3.5)}{\sqrt{30}}, \; 8.2 + \frac{2.58(3.5)}{\sqrt{30}}\right)$$
$$= (8.2 - 2.58(.64), \; 8.2 + 2.58(.64))$$
$$= (8.2 - 1.66, \; 8.2 + 1.66)$$
$$= (6.54, 9.86)$$

This finding means that the 99 per cent confidence interval is $6.54 < \mu < 9.86$. You can see that the claim that 10 pounds would be lost is rejected, in that it does not fall in the interval.

P R O B L E M

6.5.3 (a) Should the 90 per cent confidence interval contain or be contained in the 99 per cent interval?

   (b) Find the 90 per cent confidence interval for the weight-reducing problem.

EXAMPLE 3

In the last problem we determined that $\sigma_{\bar{x}} = \frac{\sigma_x}{\sqrt{n}} \approx \frac{s}{\sqrt{n}} = \frac{3.5}{\sqrt{30}} = .64$, and so $2.58\sigma_{\bar{x}} = 1.66$. That is, we found the 99 per cent interval about the point estimate $\bar{x} = 56$ to be $(56 - 1.66, 56 + 1.66)$. Suppose that instead of 1.66 we wanted a smaller interval, say 1.25, and still be 99 per cent confident. That is, we want $2.58\sigma_{\bar{x}} = 1.25$. Since $\sigma_{\bar{x}} = \frac{\sqrt{x}}{\sqrt{n}} \approx \frac{s}{\sqrt{n}}$, we can find n such that $2.58\sigma_{\bar{x}} = 1.25$. We have then

$$2.58\frac{s}{\sqrt{n}} = 1.25 \qquad \text{as } s = 3.5$$

$$2.58\frac{3.5}{\sqrt{n}} = 1.25 \qquad \text{simplifying}$$

$$\frac{9.03}{\sqrt{n}} = 1.25 \qquad \text{dividing by } 9.03$$

$$\frac{1}{\sqrt{n}} = \frac{1.25}{9.03} \qquad \text{inverting the proportion}$$

$$\sqrt{n} = \frac{9.03}{1.25} = 7.2 \qquad \text{squaring both sides}$$

$$n = 51.8 \text{---} 52 \text{ to the closest whole number}$$

Thus, in order for the 99 per cent confidence interval to have greater accuracy, we would have to increase our sample size from 30 to 52.

P R O B L E M S

6.5.4 What would be the necessary sample size n to choose if the super-market manager wanted a 95 per cent confidence interval, where $1.96\sigma_{\bar{x}} = 1.4$ and not 2.86.

6.5.5 What should the sample size n be if in the weight-reducing problem we want $1.96\sigma_{\bar{x}} = .50$ and not 1.25.

6.6 The *t*-Distribution

When we knew that the population from which we were sampling was normal, with a mean μ_x and a standard deviation σ_x, then the sample means, $\bar{x}$, chosen from random samples of size n would also have a normal distribution with mean μ_x and standard deviation $\dfrac{\sigma_x}{\sqrt{n}}$. In this situation we were able to approximate σ_x by s when we did not know σ_x. If the population was not known to be normal, then s would be a good approximation provided that n was large. In this section we shall examine what needs to be done when the distribution is normal but the size of the sample n is small.

In situations where the sample is small and the variable is normally distributed, the *t*-distribution is used in place of the standardized normal distribution. This distribution was created by a statistician named Gossett who worked for a brewery and was not allowed to publish under his own name—so he called his distribution the Student's *t*-distribution. The *t*-distribution allows us to consider the types of problems we encountered with the normal distribution in the special situation where the sample size is small and s is used in place of σ_x.

In place of the *z*-statistic, $z = \dfrac{x - \mu_{\bar{x}}}{\sigma_{\bar{x}}} = \dfrac{x - \mu}{\sigma_x/\sqrt{n}}$, we use the *t*-statistic, $t = \dfrac{x - \mu}{s/\sqrt{n}}$. Note that s replaces σ_x—it does not approximate it.

Since the *t*-distribution changes its shape depending upon n, the sample size, each *t*-distribution is associated with a number called the *degrees of freedom*. The degrees of freedom of a distribution is one less than the sample size; that is, df (degrees of freedom) $= n - 1$. Figure 6.6.1 shows *t*-distributions for df $= 1$, 9, 25, and the standard normal distribution. Notice that as the df increases, the *t*-distribution approaches the normal

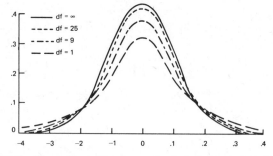

Figure 6.6.1. (From G. A. Ferguson, *Statistical Analysis in Psychology and Education.* Copyright 1959 by McGraw-Hill, Inc., and reproduced with their permission.)

distribution. In fact, when df = 25, the *t*-distribution resembles quite closely the normal distribution.

The reason that the df = $n - 1$ in particular is that the *t*-distribution is dependent upon the sample standard deviation, *s*. Recall that *s* is a function of the sum of the deviations about the mean. Thus, if you knew the mean of a set of scores, as soon as you knew the size of $n - 1$ deviations from the mean you could determine the last deviation. For example, if $\bar{x} = 4$ and $n = 5$, and $x_1 = 6$, $x_2 = 8$, $x_3 = 4$, $x_4 = 2$, and $x_5 = 0$, then $x_1 - \bar{x} = 2$, $x_2 - \bar{x} = 4$, $x_3 - \bar{x} = 0$, $x_4 - \bar{x} = -2$, and $x_5 - \bar{x} = -4$ (it must be, since the sum of the deviations from the mean must be 0). The last deviation cannot vary—it does not have the freedom to change. So by knowing $n - 1 = 4$ deviations, the last one is determined.

Since the *t*-distributions change for each df, Table IV in Appendix B shows the probability under the curve corresponding to each df. Notice that when df = ∞ (the size of the sample is infinite), the *t*-values are equivalent to the *z*-values; that is, the *t*- and normal distributions are equivalent. For example, when df = ∞, the area under the curve determined by $t_{.975}$ is 1.96. That is, $1 - .975 = .025$ is the extreme area outside the shaded region. We have found that when $z = 1.96$, there is .025 of the area from *z* to the right. For a two-sided test, any *z* that falls in the 95 per cent interval $(-1.96, 1.96)$ such that $-1.96 < z < 1.96$ would imply that the hypothesis could not be rejected. Similarly for *t*: if *t* is in the interval $(-1.96, 1.96)$ where df = ∞, the hypothesis cannot be rejected. If, however, $n = 30$, then df = $30 - 1 = 29$. We find for df = 29 that the 95 per cent interval would be $(-2.04, +2.04)$. This interval is a little wider than $(-1.96, +1.96)$. The larger the size of the sample, *n*, the better the fit between the *t* and the normal distributions.

Some applications of the *t*-distribution are explored in the next group of examples.

EXAMPLE 1

Marsha has done poorly on an exam. She thinks that the exam was really hard, and asks 10 people in the group their grades and finds that $\bar{x} = 64$ and $s = 8$. Do these data suggest that the stated class mean of $\mu = 71$ is improbable at a .05 level of significance?

For $n = 10$, $df = 10 - 1 = 9$, we shall determine the t-value associated with the data. This is found to be $t = \dfrac{x - \mu}{s/\sqrt{n}} = \dfrac{64 - 71}{8/\sqrt{10}} = \dfrac{-7}{2.53} = -2.77$. From Table IV we find that $t_{.975} = 2.26$. Since $-2.77 < -2.26$, the hypothesis that the population mean is 71 is rejected (Figure 6.6.2). However, before she

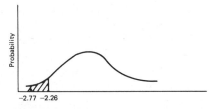

Figure 6.6.2. *t-Distribution for n = 10*

discusses the findings with her teacher, she must know if the group distribution is in fact, a normal distribution. If it is not, the t-distribution cannot be used.

PROBLEM

6.6.1 Was the test Marsha used one-tailed to the left or two-tailed?

EXAMPLE 2

A college has been using Dazzle light bulbs for years. They have found the average lifetime of bulbs to be 900 hours, with a standard deviation of 100 hours. The college is sent an advertisment for Sparkle bulbs, claiming an average life of 900 hours but at a cheaper price than Dazzle. They decide to test a sample of $n = 100$ bulbs at the .05 level of significance. They find the sample mean to be 880 hours. Can they reject the hypothesis that $H_0: \mu = 900$?

As n is large, the normal rather than the t-distribution will be used to test the claim. We then have:

1. $H_0: \mu = 900$; $H_A: \mu < 900$.
2. $\alpha \le .05$; $n = 100$.

3. Reject the null hypothesis if the z-value determined by the sample mean ($\bar{x} = 880$) and the standard deviation $\left(\sigma_{\bar{x}} = \dfrac{\sigma_x}{\sqrt{n}} = \dfrac{100}{\sqrt{100}} = 10\right)$ falls in the critical region. Since this is a one-tailed test to the left and $\alpha \le .05$, we must determine if the calculated z-value falls to the left of $z = -1.64$.

4. $z = \dfrac{\bar{x} - \mu_{\bar{x}}}{\sigma_{\bar{x}}} = \dfrac{\bar{x} - \mu_x}{\sigma_x / \sqrt{n}} = \dfrac{880 - 900}{10} = -\dfrac{20}{10} = -2.$

5. The null hypothesis is rejected at the .05 level of significance, since the observed z-value $z_0 = -2$, which is less than -1.64.

Now suppose that the sample had contained 25 light bulbs and that the population was normally distributed. In this situation we could use the t-statistic in place of the z-statistic. We then have:

1. $H_0 : \mu = 900$; $H_A : \mu < 900$.
2. $\alpha \le .05$; $n = 25$.
3. Reject the null hypothesis if the t-value determined by the sample mean ($\bar{x} = 880$), the sample standard deviation ($s = 100$), and the degrees of freedom (df $= 25 - 1 = 24$) falls in the critical region. That is, if the calculated t_0 is less than -1.71, reject the null hypothesis (Figure 6.6.3).

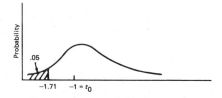

Figure 6.6.3. *t-Distribution for* $n = 25$

4. Test the null hypothesis: $t = \dfrac{\bar{x} - \mu}{s / \sqrt{n}} = \dfrac{880 - 900}{100 / \sqrt{25}} = -\dfrac{20}{20} = -1.$

5. Since $-1 > -1.71$, we cannot reject the null hypothesis at the .05 level.

How do you explain the rejection of the null hypothesis when we used the z-statistic but not when we used the t-statistic?

EXAMPLE 3

The following is a test to determine if there is a significant difference between the null hypothesis that $\mu = 200$, and the sample findings of $\bar{x} = 196$, $s = 8$, and $n = 15$. Fill in the missing steps.

1. $H_0 : \mu = 200$; $H_A : \mu < 200$.
2. $\alpha \le .05$; $n = 15$.

P R O B L E M

6.6.2 Suppose the situation in Example 3 is tested again but that this time $n = 6$. As such, df $= 6 - 1 = 5$. If the calculated t_0 is less than -2.02, the null hypothesis is rejected. Is it? Sketch the t-distribution for $n = 6$—include the critical region.

The same approach that is used to determine confidence intervals for population means based on sample findings from a normal distribution can be used for confidence intervals for the t-distribution with one exception. Recall that if $\alpha \le .05$, the critical region was $z = 1.96$ for a two-tailed test and $z = 1.64$ for a one-tailed test. For a normal distribution test this is the case regardless of the size of the sample n. Now that we are considering small samples, the value of t is determined by α and also by the degrees of freedom. For example, if:

1. df $= 8$, $\alpha \le .05$, the 95 per cent confidence interval is $-2.31 < \mu < 2.31$.
2. df $= 12$, $\alpha \le .05$, the 95 per cent confidence interval is $-2.18 < \mu < 2.18$. We shall determine a 95 per cent confidence interval using a t-distribution for a population mean, μ, in the next example.

EXAMPLE 4

Suppose that a random sample of $n = 16$ students is chosen from a school population. If their mean score on a nationwide intelligence test is 106, with a standard deviation of 10, what is a 95 per cent confidence interval for the mean intelligence of the student body?

We want to determine the lower and upper bounds of an interval that we feel 95 per cent confident contains the population mean:

$$\bar{x} - t_{.975} \frac{s}{\sqrt{n}} < \mu < \bar{x} + t_{.975} \frac{s}{\sqrt{n}}$$

As $\bar{x} = 106$, $s = 10$, and df $= 16 - 1 = 15$, we have that

$$106 - 2.13 \frac{10}{\sqrt{16}} < \mu < 106 + 2.13 \frac{10}{\sqrt{16}}$$

$$= 106 - 5.325 < \mu < 106 + 5.325$$

$$= 100.675 < \mu < 111.325$$

If $n = 25$ and not 16, then when $\bar{x} = 106$ and $x = 10$, the 95 per cent confidence interval would be affected by df $= 25 - 1 = 24$ and not 15:

$$106 - 2.06\frac{10}{\sqrt{25}} < \mu < 106 + 2.06\frac{10}{\sqrt{25}}$$

$$106 - 2.06(2) < \mu < 106 + 2.06(2)$$

$$100.88 < \mu < 111.12$$

This 95 per cent confidence interval is narrower than the one we determined when $n = 16$. As n increases, the interval narrows as it approaches the population mean.

EXAMPLE 5

Suppose that a firm has received a sample of five typewriter ribbons of a new material that the manufacturer claims lasts longer than the standard ribbon. The average life of a standard ribbon is 40 days in this firm. The new ribbons are tried on randomly selected typewriters, and the average life of these ribbons is determined to be 51 days, with a standard deviation of 15 days.
  (a) Test the hypothesis that the new ribbons are not an improvement over the standard ribbons.
  (b) Construct the 95 per cent confidence limits.

Solution
  (a) The tests in the hypotheses test as as follows:
    1. $H_0: \mu = 40$; $H_A: \nu > 40$.
    2. $\alpha \le .05$; $n = 5$.
    3. Since $\bar{x} = 51$ and $s = 15$, then $\sigma_{\bar{x}} = 15/\sqrt{5} = 6.72$. We find that $t_0 = (51 - 40)/6.72 = 1.34$. The t-distribution for $n = 5$ and $\alpha = .05$ determines that when $t > 2.13$ there is 5 per cent of the distribution to the right of $t_{.95} = 2.13$.
    4. Since $t_0 = 1.34 < 2.13$, we cannot reject the hypothesis; that is, there is no significant difference between the two types of ribbons. Did you think that the hypothesis would be rejected?
  (b) $\bar{x} - t_{.975} \cdot \sigma_{\bar{x}} \le \mu \le \bar{x} + t_{.975}\alpha_{\bar{x}}$ is the interval we wish to determine. Since $\bar{x} = 51$ and $t_{.975} = 2.78$ when df $= 4$ and $\sigma_{\bar{x}} = 6.72$, we have $51 - 2.78(6.72) \le \mu \le 51 + 2.78(6.72)$, which simplifies to $32.3 \le \mu \le 69.7$. Notice that this interval is quite wide. This is due to the small sample size.

PROBLEMS

6.6.3 (a) Determine the 90 and 99 per cent confidence intervals for the mean of the population if $\bar{x} = 210$, $s = 25$, and $n = 25$.
  (b) How would you explain the relationship between the width of the confidence interval and the per cent of confidence?

6.6.4 Suppose that $n = 25$ new typewriter ribbons were tested and the mean and standard deviation were as given in Example 5. Determine (a) and (b) again.

6.7 Summary

In this chapter we examined some of the characteristics and applications of the normal distribution. The mean, median, and mode were seen to be located at the same center point on the horizontal axis of this symmetric, bell-shaped distribution. To determine the location and shape of a normal distribution, the mean and standard deviation of the distribution must be known. We saw that if the normal distribution was thin and tall (having a small standard deviation) or short and squat (having a large standard deviation), there would be approximately 68.3 per cent of the total area within 1 standard deviation of the mean, 95.5 per cent within 2 standard deviations of the mean, and 99.7 per cent within 3 standard deviations of the mean.

To make wide use of the normal distribution, the distribution was standardized so that the mean, $\mu = 0$, and the standard deviation, $\sigma = 1$. In this way we could convert and compare scores from different normal populations. The normal distribution was also used to simplify problems involving a binomial distribution. The approximation of the normal to the binomial distribution became better as n increased, and provided a good fit as soon as $np > 5$ if p was smaller than q; or $nq > 5$ if q was smaller than p.

We tested hypotheses made about normal populations using the same form as the hypothesis test for the binomial distribution. If $\alpha \leq .05$, a two-tailed test required that if $z > 1.96$ or $z < -1.96$, the hypothesis was rejected. For a one-tailed test and $\alpha \leq .05$, *if $z > 1.64$* (one-tailed to the right) or if $z < -1.64$ (one-tailed to the left), the hypothesis was rejected. This was the case since the interval $(-1.96, +1.96)$ contains 95 per cent of the distribution, and so do $(-1.64, +\infty)$ and $(-\infty, +1.64)$. In order to test hypotheses about the population mean based on sample findings, we examined the sampling distribution. It was found that if all possible samples of size n are taken from a normal population, then the population mean, μ_x, and the mean of all the sample means, $\mu_{\bar{x}}$, would be equal. In addition, the relationship between the population standard deviation σ_x and the standard deviation of the sample means would be that $\sigma_{\bar{x}} = \dfrac{\sigma_x}{\sqrt{n}}$.

What is remarkable is that even if the population distribution is not normal, the sampling distribution of $\bar{x}$ would be approximately normal, and the approximation would be better as n increased.

Confidence intervals for population means were created similar to confidence intervals for population proportions. The general form of the confidence interval was found to be $(\bar{x} - z_0 \cdot \sigma_{\bar{x}}, \bar{x} + z_0 \cdot \sigma_{\bar{x}})$, where z_0 was determined by the degree of confidence required. If a 95 per cent confidence interval was desired, $z_0 = 1.96$. The sample standard deviation, s, was used as an approximation to σ_x when σ_x was unknown and n was large. And, since $\sigma_{\bar{x}} = \dfrac{\sigma_x}{\sqrt{n}}$, the 95 per cent confidence interval appears as

$\left(\bar{x} - 1.96\dfrac{s}{\sqrt{n}}, \bar{x} + 1.96\dfrac{s}{\sqrt{n}} \right)$. For any sample chosen, we would obtain a sample mean, $\bar{x}$, and a sample standard deviation, s. When we use $z_0 = 1.96$, we can be 95 per cent confident that the particular confidence interval contains the population mean. In 5 per cent of these confidence intervals, a sample mean would have been determined which is relatively far away from the population mean, and thus the confidence interval would fail to contain the population mean.

The t-distribution was introduced when only a small sample could be taken from the normal population. When the sample is large, the substitution of s for σ is acceptable. However, when it is small, the substitution is not appropriate. The t-distribution uses only the sample mean and standard deviation and does not require substituting the sample standard deviation for the population standard deviation. Each t-distribution is determined by the size of the sample. As n increases, the t-distribution moves closer to the normal distribution. Even for $n = 25$, the approximation is quite good. In order to use the t-statistic in place of the z-statistic, the population distribution must be normal. The sample size n is reflected in the degrees of freedom, $n - 1$, which determines the shape of the t-distribution. In testing hypotheses based on small samples, we used

$$t = \frac{\bar{x} - \mu}{s/\sqrt{n}} \quad \text{in place of} \quad z = \frac{\bar{x} - \mu}{\sigma/\sqrt{n}}$$

The confidence interval took the form of $\left(\bar{x} - t_0 \cdot \dfrac{s}{\sqrt{n}}, x + t_0 \cdot \dfrac{s}{\sqrt{n}} \right)$. The value of t_0 was determined not only by the degree of confidence, but also by the degrees of freedom. A 95 per cent confidence interval when df $= 12$ set $t_{.975} = 2.18$, while if df $= 24$, $t_{.975} = 2.06$.

6.8 Review of Formulas

  I. The normal distribution (Figure 6.8.1)
    A. Area under the curve
      1. 68+ per cent of the area within 1σ of the mean.
      2. 95+ per cent of the area within 2σ of the mean.
      3. 99+ per cent of the area within 3σ of the mean.

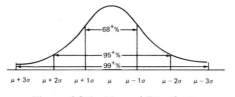

Figure 6.8.1. *Normal Distribution*

B. Conversion formula

To convert from a set of x-scores taken from a normal distribution to a set of z-scores associated with the standardized normal distribution ($\mu = 0; \sigma = 1$), the following conversion is done: $x \rightarrow z = \dfrac{x - \mu}{\sigma}$.

C. Normal approximation to the binomial

Since the binomial distribution contains discrete data and the normal distribution continuous data, the conversion from an x-score to a z-score requires that the value of x first be decreased by .5, before converting to a z-score, if $P(x \geq \underline{\quad})$ is required. For example, the binomial probability value of $P(x \geq 6)$ would be converted to the normal approximation of $P(x \geq 5.5)$. If $P(x \leq \underline{\quad})$ is required, the x-score is first increased by .5 before converting to a z-score. For example, the binomial probability value of $P(x \leq 6)$ would be converted to the normal approximation of $P(x \leq 6.5)$.

D. Hypothesis testing
  1. Two-tailed test
     a. $\alpha \leq .05$; if z falls outside $(-1.96, +1.96)$, the hypothesis is rejected.
     b. $\alpha \leq .10$; if z falls outside $(-1.64, +1.64)$, the hypothesis is rejected.
  2. One-tailed test to the right
     a. $\alpha \leq .05$; if z is greater than 1.64, the hypothesis is rejected.
  3. One-tailed test to the left
     a. $\alpha \leq .05$; if z is less than -1.64, the hypothesis is rejected.

E. Sampling distribution of $\bar{x}$
  1. If the original distribution is normal with mean μ_x and standard deviation σ_x, then the sampling distribution of $\bar{x}$ is normal with mean $\mu_{\bar{x}} = \mu_x$ and standard deviation $\sigma_{\bar{x}} = \dfrac{\sigma_x}{\sqrt{n}}$.
  2. If the original distribution is not normal with mean μ_x and standard deviation σ_x, then the sampling distribution of $\bar{x}$ will

be approximately normal with mean $\mu_{\bar{x}} = \mu_x$ and standard

deviation $\sigma_{\bar{x}} = \dfrac{\sigma_x}{\sqrt{n}}$.

3. The conversion of an x-score to a z-score for a sampling

distribution is $x \to z = \dfrac{x - \mu_{\bar{x}}}{\sigma_{\bar{x}}} = \dfrac{x - \mu_x}{\sigma_x / \sqrt{n}}$.

F. Estimation
   We have considered two estimates of the population mean:
   1. Point estimate: $\bar{x}$, the sample mean.

   2. Interval estimate: $\left(\bar{x} - z_0 \cdot \dfrac{\sigma_x}{\sqrt{n}}, \ \bar{x} + z_0 \cdot \dfrac{\sigma_x}{\sqrt{n}} \right)$, where $z_0 = 1.96$

   for the 95 per cent confidence interval, and $z_0 = 2.58$ for the
   99 per cent confidence interval. If n is large, then σ_x may be
   approximated by s.

II. *t*-Distribution
   The *t*-distribution can be used when n is small and the original
   distribution is assumed normal. Since n is small, the approximation
   of σ_x by s is not used. The *t*-distribution conversion of an x-score to
   a *t*-score does not depend on σ_x. The sample standard deviation, s, is
   used in place of σ_x.

   A. The conversion formula is $x \to t = \dfrac{x - \mu_x}{s / \sqrt{n}}$, with df $= n - 1$.

   B. Hypothesis testing
      The hypothesis test for a *t*-distribution depends upon the level of
      significance, α, and the degrees of freedom, unlike the hypothesis
      test for a normal distribution, which depends only upon the level
      of significance.
      1. Two-tailed test
         a. $\alpha \leq .05$, and df $= n - 1 = 15$; if t falls outside $(-2.13, +2.13)$,
            the hypothesis is rejected.
         b. $\alpha \leq .05$, and df $= n - 1 = 20$; if t falls outside $(-2.09, +2.09)$,
            the hypothesis is rejected.
      2. One-tailed test to the right
         a. $\alpha \leq .05$, and df $= 15$; if $t > 1.75$, the hypothesis is rejected.
      3. One-tailed test to the left
         a. $\alpha \leq .05$, and df $= 20$; if $t < -1.72$, the hypothesis is rejected.

   C. Estimation
      There are two estimates of the population mean:
      1. Point estimate: $\bar{x}$, the sample mean.

      2. Interval estimate: $\left(\bar{x} - t_0 \cdot \dfrac{s}{\sqrt{n}}, \ \bar{x} + t_0 \cdot \dfrac{s}{\sqrt{n}} \right)$, where if df $= 12$,

      $t_0 = t_{.975} = 2.18$ for the 95 per cent confidence interval, and if
      df $= 25$, $t_{.975} = 2.06$, for the 95 per cent confidence interval.

6.9 Problem Set

6.9.1 The following statistics are of five groups, each of whose distribution was approximately normal and whose sample size was $n = 50$:

Group	$\bar{x}$	s
1	50	10
2	50	12
3	65	10
4	70	3
5	70	5

Which group do you think:
(a) Would have the highest individual score?
(b) Would have the lowest individual score?
(c) Has the largest range?
(d) Has the smallest range?
(e) Most closely resembles the entire group of $n = 250$?

6.9.2 If the mean lifetime of a company's light bulbs is 1,600 hours with a standard deviation of 100 hours, what is the interval that is 1σ of the mean; 2σ of the mean? If you knew that the distribution of bulb lifetimes was normally distributed, what additional statements could you make?

6.9.3 Suppose that the bulb lifetime distribution was normally distributed. What per cent of the distribution lies in the intervals (a) (1,600, 1,700); (b) (1,600, 1,800); (c) (1,500, 1,700); (d) (1,450, 1,650)?

6.9.4 Convert the following x-scores to standardized z-scores using the lifetime bulb distribution: (a) 1,600; (b) 1,700; (c) 1,750; (d) 1,900.

6.9.5 If the bulb lifetime distribution was normally distributed, find the per cent of hours that would be expected for the bulbs to burn:
(a) Between 1,600 and 1,650.
(b) At most 1,650.
(c) At least 1,700.
(d) Between 1,700 and 1,900.
(e) Between 1,400 and 1,550.
(f) Between 1,450 and 1,750.

6.9.6 Compare the exact binomial probability value to the approximate normal probability values in the following situation. A stop-smok-

ing program claims that it was 70 per cent successful. Find the
probability that in a sample of $n = 20$ people:
(a) At least 14 stop.
(b) At least 17 stop.
(c) Between 15 and 18 stop.
(d) At most 10 stop.
(e) Between 10 and 16 stop.

6.9.7 Find the probability of obtaining between 40 and 60 (inclusive)
heads in tossing a fair coin 100 times.

6.9.8 What would be the critical region and acceptance region if a fair
coin was flipped 64 times and $\alpha \leq .05$; $\alpha \leq .10$?

6.9.9 If 160 of 200 people were satisfied with a new product, test the
hypothesis that $P(\text{satisfied}) = .90$ against the alternative hypothesis
that $P(\text{satisfied}) < .90$, at a .05 level of significance.

6.9.10 Find the least number of satisfied people necessary in order not to
reject the hypothesis that $P(\text{satisfied}) \geq .90$ at a .01 level of
significance.

6.9.11 The mean lifetime of a new battery is claimed to be 800 hours. If a
sample of 100 batteries determines a sample mean of 790 hours and
a standard deviation of 55 hours, test the hypothesis $\mu = 800$
against $\mu \neq 800$, for $\alpha \leq .05$; for .10.

6.9.12 The following table relates the level of significance, α, to the critical
z-values for a one- and a two-tailed test. Why are the values 1.64
and 2.58 repeated? Why is each value expressed as "$\pm$"?

	α			
	.10	.05	.01	.005
z_0 for one-tailed	± 1.28	± 1.64	± 2.33	± 2.58
z_0 for two-tailed	± 1.64	± 1.96	± 2.58	± 2.81

6.9.13 If IQs are normally distributed with $\mu = 100$ and $\sigma = 10$, what is the
probability that the average IQ of 25 people chosen at random is at
most 107?

6.9.14 Has television hurt sports attendance? If the average person used to
attend on the average 15 sports shows per year before television
presented them, test the hypothesis that $\mu = 15$ versus $\mu < 15$, if a
sample of $n = 200$ fans yields a sample mean of 12 and a standard
deviation of 8, at a .05 level of significance.

6.9.15 A city campaign to raise funds resulted in $100,000 being collected by 5,000 volunteers. While a point estimate of the average collection can be determined easily, the sponsors want a clearer understanding of the average collection. For example, at this point the sponsors do not know if 1 volunteer collected $98,000 and the other 4,999 volunteers collected $2,000. A sample is taken of 100 volunteers. it is found that the average volunteer received 10.2 contributions with a standard deviation of 5.
(a) What is the point estimate of the mean?
(b) What is the 95 per cent confidence interval for the number of contributors per volunteer?
(c) What is the 95 per cent confidence interval for the number of contributions?

6.9.16 Test the null hypothesis $H_0 : \mu = 90$ against the alternative hypothesis, $H_A : \mu \neq 90$ with $\alpha \leq .10$, if $\bar{x} = 85$, $s = 10$, and $n = 20$.

6.9.17 A garage owner finds that the average customer pays $30 for repairs and that the cost per customer is approximately normally distributed. She is concerned that the high cost of living has cut down on her profit. She chooses a random sample of 10 customers over the following weeks and the money they spent (dollars): 5, 15, 20, 20, 25, 25, 30, 35, 35, and 40. Test the hypothesis that there is no difference between the assumed mean and the sample mean. Use a one-tailed test to the left and a level of significance of .025.

6.9.18 Calculate the 95 per cent confidence interval for the mean of the garage owner.

6.9.19 Which of the following distributions is (are) most likely to be approximately normal?
(a) The distances a baseball could be thrown by all the boys in the tenth grade at a high school.
(b) The Scholastic Aptitude test scores of high school freshman.
(c) The distances a baseball could be thrown by all the students in the tenth grade at a high school.
(d) The heights of 1,000 poplars planted at the same time 2 years ago.

6.9.20 In a normal distribution, what is the z-score for the 25th percentile?

6.9.21 In a normal distribution, a z-score of $+2.39$ has what percentile rank?

6.9.22 The scores of children on the Wechsler Intelligence Scale for Children are approximately normal with $\mu = 100$ and $\sigma = 15$. What percentage of the children can be expected to score between 85 and 110 on the WISC?

6.9.23 On the WISC, what scores will occur only 2 per cent of the time?

6.9.24 The manager of a flea market knows that from past experience his attendance distribution is approximately normal with a mean of 372 and a standard deviation of 32. If his break-even point is an attendance of 290 people, how often does he lose money?

6.9.25 Pooky, a cat, belongs to an age group in which the average weight is 10 pounds with a standard deviation of 2.5 pounds, and Girl, another cat, belongs to an age group in which the average weight is 13 pounds with a standard deviation of 4 pounds. If Pooky weighs 14.5 pounds and Girl weighs 19 pounds, which of the two is more overweight with respect to his or her own age group?

6.9.26 On an examination the average grade was 72 and the standard deviation was 6. If 10 per cent of the class are given A's, and the grades are curved to fit a normal distribution, what is the lowest possible score of A and the highest possible score of B?

6.9.27 Luis rides his bike to his accounting office every day, weather permitting. He finds that his average time is 29 minutes with a standard deviation of 3 minutes. If we assume that the distribution of trip times to be normally distributed, find:
(a) The probability that his trip will take at least 35 minutes.
(b) The percentage of time he arrives after 9:00 A.M. if he leaves his house at 8:30, 8:25, and 8:15?
(c) The probability that he arrives after his secretary does if she arrives between 8:50 and 9:00 A.M. and he leaves his house at 8:20 A.M.?

6.9.28 The average life of a certain watch is 10 years, with a standard deviation of 2 years. The manufacturer will replace free all watches that fail under guarantee; the life-time distribution of the watches is approximately normal. If he is willing to replace only 4 per cent of the watches that fail, how long a guarantee should he give?

Chapter 7
Distribution-Free Tests

7.1 Introduction

In Chapters 5 and 6 we tested hypotheses where the data were assumed to have come from either binomial or normal populations. We had to be able to demonstrate or reasonably assume that the populations satisfied either the binomial or normal distribution assumptions, if we were to use them. This chapter will examine some valuable statistical tests that do not make any assumptions about the types of populations from which the sample data came, and therefore have wider applications. While these tests nevertheless do have assumptions that must be satisfied before they can be used, they are not as demanding as the binomial and normal tests. We shall examine four statistical tests that provide us with additional ways of testing hypotheses:

1. Chi-square (read: ki-square) test for goodness of fit, where we compare if what we observed "fits" what we expected; for example, can we expect that the distribution of types of records sold today are the same as they were 10 years ago?
2. Chi-square test for independence, where we test if two variables are independent of, or dependent on each other; for example, is a person's favorite type of television show independent of, or dependent on, the person's age?
3. Sign test, where we test if members of one group who were originally matched to members of a second group are significantly different in some activity as a result of different treatments; for example, suppose that two groups of students were matched on weight, that is, each member of group A weighs the same as each member of group B. Does the fact that a special exercise program is being given to group A create a significant decrease in weight as compared to group B?
4. U-test, where we test if two independent samples are significantly different from one another as a result of the effects of different treatments; for example, do two groups that are chosen independently, that is, they are not matched on any specific characteristics, react significantly differently to different types of wake-up approaches?

If you plan to take a survey, the chi-square tests could be of help to you in analyzing your data. For example, the chi-square goodness-of-fit test could be used in the situation where you asked a question that had three response categories and you had guessed that the distribution of responses would be in the ratio $1:2:1$. You could use this test to determine if there was a significant difference between your findings and your guess. The chi-square independence test could be of value in determining if different groups, depending upon religion, sex, education, and so on, responded significantly differently to particular questions. The sign test could be used in a survey of growth patterns. For example, the test could be of help in determining if the amount of growth is significantly different for two groups of bean sprouts that are matched on initial height when one group is given high light and the other low light. As you read the discussion of the U-test, see if you can determine how this test might be of help to you in your survey.

7.2 Chi-Square (χ^2) Goodness-of-Fit Test

The χ^2 test provides us with a method to make statistical judgments as to whether what we expected to occur agrees with what we observed to occur. Suppose we wanted to test the hypothesis that Reggie and Jackie spend their free time now as they used to by: seeing movies (20 per cent), seeing plays (10 per cent), watching television (50 per cent), and listening to music (20 per cent). We observe that in the next 50 times they have free time together they spend that time by seeing movies (11 times), see plays (2 times), watching television (31 times), and listening to music (6 times). To test this hypothesis using a binomial test, we would have to create only two categories (e.g., movies or no movies). But doing this wouldn't give us a clear picture of how they spend their total free time. In order to take into consideration all these categories, we need a new statistical test, a test that:

1. Allows us to make judgments, based on sample data, without knowing the population from which the sample came.
2. Allows us to consider more than two types or categories of outcomes.

The chi-square test allows us to test hypotheses about which we make no assumptions about the populations. However, there are assumptions that must be met in order to use this test:

1. The outcomes are independent of each other.
2. Every observation falls into one and only one category.
3. The expected frequency in each category must have at least five occurrences.

This test enables the tester to examine if there is, in fact, a significant difference between what was actually observed and what was expected.

In the problem mentioned earlier we found that the observed distribution was $o_1 = 11$ (movies), $o_2 = 2$ (plays), $o_3 = 31$ (television), and $o_4 = 6$ (music). That is, o_1 is the number of observed occurrences in the first category (movies); o_2 is the number of observed occurrences in the second category (plays); and so on. What are e_1, e_2, e_3, and e_4—the number of expected occurrences for each category? Since they usually see movies 20 per cent of the time, then, as $N = 50$, we should expect they would see $E(x) = Np = 50(.20) = 10$ movies; that is, $e_1 = 10$. And $e_2 = 50(.10) = 5$; $e_3 = 25$; and $e_4 = 10$. We summarize the information in tabular form in Table 7.2.1. This representation does not make clear what we want to compare: namely, o_1 to e_1, o_2 to e_2, and so on. Therefore, we shall rewrite the table as shown in Table 7.2.2.

Table 7.2.1. Frequency of Observed and Expected Occurrences

o_1	o_2	o_3	o_4	e_1	e_2	e_3	e_4
11	2	31	6	10	5	25	10

Table 7.2.2. Frequency of Observed and Expected Occurrences

	Category			
Outcomes	1	2	3	4
Observed	11	2	31	6
Expected	10	5	25	10

The question that is to be answered is: Is there a significant difference between what was observed and what was expected? Notice that the sum of the observed, $o_1 + o_2 + o_3 + o_4 = 50$, must equal the sum of the expected, $e_1 + e_2 + e_3 + e_4 = 50$—this must always be the case. We can see that there are, in fact, differences; $o_1 \neq e_1$, $o_2 \neq e_2$, and so on. But the difference could be due to sampling variation and not to any significant change in Reggie and Jackie's habits. We want to determine if the difference is significant— that is, the difference great enough that it is probably not due to chance? To determine the difference, we might try to find the sum of the differences: $(o_1 - e_1) + (o_2 - e_2) + (o_3 - e_3) + (o_4 - e_4)$. If this is done, the sum is found to be zero, since $(11 - 10) + (2 - 5) + (13 - 25) + (6 - 10) = 0$. This will always be the case. See if you can prove that $(o_1 - e_1) + (o_2 - e_2) + \cdots + (o_n - e_n) = 0$, where there are n categories. (HINT: Rearrange the terms and use the fact that is mentioned in the paragraph before this one.) The chi-square statistic that is used is

$$\chi^2 = \frac{(o_1 - e_1)^2}{e_1} + \frac{(o_2 - e_2)^2}{e_2} + \cdots + \frac{(o_n - e_n)^2}{e_n}$$

That is, each difference is squared (which eliminates negative differences) and each squared difference is divided by the corresponding expected frequency. The division of each squared difference by the expected frequency is done to show the squared difference *relative* to the expected frequency. For example, suppose that $o_1 = 503$ and $e_1 = 500$. Then $(o_1 - e_1)^2 = (503 - 500)^2 = 9$. But if $o_1 = 4$ and $o_2 = 1$, $(o_1 - e_1)^2 = (4 - 1)^2 = 9$. Is it reasonable to write both squared differences as 9? Since 4 is four times 1 and 503 is 1.006 times 500, we want to represent the "relative closeness" in the calculation. We use $\dfrac{(o_1 - e_1)^2}{e_1}$. And so when $o_1 = 503$ and $e_1 = 500$,

$\dfrac{(o_1 - e_1)^2}{e_1} = \dfrac{9}{500} = .018$, while when $o_1 = 4$ and $e_1 = 1$, $\dfrac{(o_1 - e_1)^2}{e_1} = \dfrac{9}{1} = 9$. In this manner we can see that the observed frequency was close to the expected frequency in the first case and not in the second.

The data in Table 7.2.2 determine that

$$\chi^2 = \frac{(11 - 10)^2}{10} + \frac{(2 - 5)^2}{5} + \frac{(31 - 25)^2}{25} - \frac{(6 - 10)^2}{10}$$

$$= \frac{1}{10} + \frac{9}{5} + \frac{36}{25} + \frac{16}{10} = \frac{5 + 45 + 72 + 80}{50} = \frac{202}{50} = 4.04$$

The question we have to answer, as we did in our earlier hypothesis testing is: Is $\chi^2 = 4.04$ large enough to say that the difference is significant? To answer this question, we need to examine Table V in Appendix B.

For example, for df $= 5$, $\chi^2_{.95} = 11.1$. This means that when there are $n = 6$ categories, so df $= n - 1 = 5$, a χ^2 value as much as 11.1 will occur 95 per cent of the time. Any $\chi^2 > 11.1$ will occur at most 5 per cent of the time. As such, if we had df $= 5$ and a χ^2 value greater than 11.1, we would reject the hypothesis that there is no difference between what we observed and what we expected, since a value of χ^2 larger than 11.1 would occur by chance at most 5 per cent of the time. When the df $= 10$ and $\chi^2_{.95} = 18.3$, there are $n = 11$ categories, and a $\chi^2 > 18.3$ will occur at most 5 per cent of the time. Can you think of a reason why we are using $\chi^2_{.95}$ and not $\chi^2_{.50}$ or $\chi^2_{.10}$? If you reasoned that a $\chi^2_{.95}$ determines a .05 critical region, that is, $\alpha \le .05$, you are absolutely correct. Several χ^2 distributions are given in Figure 7.2.1; note that the df determines different χ^2 distributions.

The reason that df $= n - 1$ in particular is that there are only $n - 1$ categories with frequencies which can vary. Table 7.2.2 contains $n = 4$ categories with a total frequency of $N = 50$. As soon as the frequencies in any of the $n - 1 = 3$ categories are known, the frequency in the last category can be calculated; that is, it cannot vary, it does not have the freedom to change. For example, since $N = 50$ and $o_1 + o_2 + o_3 = 44$, then $o_4 = 50 - 44 = 6$. In the problem we were considering, $\chi^2 = 4.04$ and df $= n - 1 = 4 - 1 = 3$. We find in Table V that $\chi^2_{.95} = 7.81$ when df $= 3$. Since $4.04 < 7.81$, we cannot reject the hypothesis that Reggie and Jackie spend their

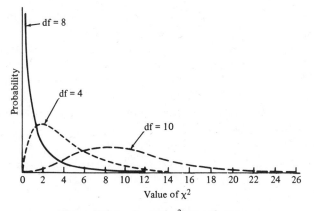

Figure 7.2.1. *Several χ^2 Distributions*

free time as usual. That is, the calculated χ^2 is small enough to claim that the difference, as measured by χ^2, is due to chance, and not to a real difference between the expected and the observed.

In the problem that we just considered, the question was whether the assumed values (e_1, e_2, e_3, e_4) was a good fit for the observed values (o_1, o_2, o_3, o_4). A good fit is claimed if the value of χ^2 is small. That is, the difference between the observed and the expected values is due to chance, not to a misunderstanding of the situation. Consider the following situation: A teacher believes that there is not one "best" teaching method that is favored by most students. To test his hypothesis, he collects a random sample of 100 students and demonstrates five different teaching methods to them. He then asks them to choose the one they found to be the best. The expected and observed outcomes are listed in Table 7.2.3. The fact that $e_1 = e_2 = \cdots = e_5 = 20$ is based on the assumption that different students favor different teaching methods. Notice that the sum of the observed frequencies, $o_1 + o_2 + o_3 + o_4 + o_5$, is equal to the sum of the expected frequencies, $e_1 + e_2 + e_3 + e_4 + e_5$, and that each response falls into one and only one category and is independent of the other responses.

Table 7.2.3

Method	Frequency of expected outcomes	Frequency of observed outcomes
A	20	14
B	20	21
C	20	23
D	20	17
E	20	25
	100	100

The χ^2 hypothesis test can be put in the form that we have been using:
1. Create the null and alternative hypotheses:

    H_0: There is no significant difference in the teaching methods.

    H_A: There is a significant difference in the teaching methods.
2. Level of significance—$\alpha \le .05$; number of categories is $n = 5$; thus df $= 5 - 1 = 4$.
3. Reject the null hypothesis if the calculated $\chi^2 > \chi^2_{.95}$ for df $= 4$; that is, if $\chi^2 > 9.49$ (from Table V).
4. Test the null hypothesis with sample data:

$$\chi^2 = \frac{(14-20)^2}{20} + \frac{(21-20)^2}{20} + \frac{(23-20)^2}{20} + \frac{(17-20)^2}{20} + \frac{(25-20)^2}{20}$$

$$= \frac{36}{20} + \frac{1}{20} + \frac{9}{20} + \frac{9}{20} + \frac{25}{20} = \frac{70}{20} = 3.5$$

5. Since the calculated $\chi^2 = 3.5$ and $3.5 < 9.49$, the null hypothesis cannot be rejected at the .05 level of significance.

Expected frequencies, besides being based on personal assumptions or past experience, can be based on a scientific principle. For example, if two pink flowers were crossed and each contained a red and a white gene (neither was dominant), over the long run we would expect to obtain an equal number of red and white flowers and twice as many pink as red or white. Suppose that, after numerous cross plantings, we had obtained 65 red, 114 pink, and 57 white; can we claim that neither gene is dominant?

We know that $o_1 = 65$, $o_2 = 114$, and $o_3 = 57$—but what are the values of e_1, e_2, and e_3? Since we expect that 25 per cent will be red, 25 per cent white, and 50 per cent pink, when we find n, we can find $E(x)$, the expected number of each. The sum of the observed, $o_1 + o_2 + o_3 = 65 + 114 + 57 = 236 = N$. If we let $e_1 =$ expected number of red, $e_2 =$ expected number of white, and $e_3 =$ expected number of pink, we have $e_1 = (.25)(236) = 59$, $e_2 = (.25)(236) = 59$, and $e_3 = (.50)(236) = 118$. Table 7.2.4 contains the

Table 7.2.4

Outcome	Category 1	Category 2	Category 3
Observed	65	57	114
Expected	59	59	118

results. The null hypothesis is: Neither red nor white is a dominant gene. As there are $n = 3$ categories, there are $3 - 1 = 2$ df. Using a level of significance of .05, we shall reject the null hypothesis if $\chi^2 > 5.99$ (Table V).

We find

$$\chi^2 = \frac{(65-59)^2}{59} + \frac{(57-59)^2}{59} + \frac{(114-118)^2}{118} = \frac{96}{118}$$

Since $\frac{96}{118} < 5.99$, we cannot reject the hypothesis.

The following situation considers the problem of what to do when one of the expected categories has a frequency of less than 5. Larry and Helge's Pizzeria usually sells regular pizzas, sausage pizzas, mushroom pizzas, and meatball pizzas in the ratio $5:2:2:1$. Owing to a rise in costs, they had to raise the price of their pizzas, especially the sausage pizzas. The next 40 orders produced the sales distribution shown in Table 7.2.5. Can they say that sales are as usual? The ratio of $5:2:2:1$ implies that $5x + 2x + 2x + 1x = 40$, and $x = 4$. The observed and expected frequencies are listed in Table 7.2.6. The hypothesis testing must be held up, as one of the expected categories has a frequency less than 5. One of two alternatives should be chosen:

Table 7.2.5

Category	Frequency
Regular	26
Sausage	4
Mushroom	5
Meatball	5

Table 7.2.6

Outcome	Category			
	1	2	3	4
Observed	26	4	5	5
Expected	20	8	8	4

1. Combine two categories so that the sum is at least 5. Sometimes this alternative causes problems in that, by grouping, two categories are put together that really should not be. For example, if there were 4 choices for president, grouping the votes for two candidates together would really make no sense.
2. Use Yates' correction formula (a formula used in place of the χ^2

formula when the frequency of any expected category is less than 5 or
if df $= 1$):

$$\chi^2 = \frac{(|o_1 - e_1| - .5)^2}{e_1} + \cdots + \frac{(|o_n - e_n| - .5)^2}{e_n}$$

where $|o_i - e_i|$ represents the difference of the ith observed and
expected frequencies and the sign of the difference is dropped.

Since Larry and Helge are concerned about their pizza sales, we shall
combine the last two categories. The new presentation is (Table 7.2.7):

Table 7.2.7

Outcome	Category		
	1	2	3
Observed	26	4	10
Expected	20	8	12

1. H_0: Sales are still in the ratio $5:2:3$ $(2+1)$.
   H_A: Sales are not in the usual ratio.
2. $\alpha \le .10$; df $= 3 - 1 = 2$.
3. Reject the null hypothesis if $\chi^2 > 4.61(\chi^2_{.90})$.
4. $\chi^2 = \dfrac{(26-20)^2}{20} + \dfrac{(4-8)^2}{8} + \dfrac{(10-12)^2}{20} = 4.13.$
5. As $4.13 < 4.61$, we cannot reject the hypothesis that sales are as
   usual.

When there are only two categories, then $n = 2$ and df $= 2 - 1 = 1$. In the
case that df $= 1$, we must use Yates' correction formula to adjust for the
small degrees of freedom. The following example uses the correction
formula. A random sample of $N = 50$ people were asked if they were for or
against abortion. If 29 favor abortion and 21 are against, can we conclude
that the community is split 50–50? Table 7.2.8 organizes the information.

Table 7.2.8

	For	Against
Observed	29	21
Expected	25	25

1. H_0: The community is evenly split on the question of abortion.
   H_A: The community is not evenly split on this question.

2. $\alpha \leq .10$; df $= 2 - 1 = 1$.
3. Reject the null hypothesis $\chi^2 > 2.71 (\chi^2_{.90})$.
4. Using Yates' correction formula:

$$\chi^2 = \frac{(|29-25|-.5)^2}{25} + \frac{(|21-25|-.5)^2}{25} = \frac{(3.5)^2}{25} + \frac{(3.5)^2}{25} = .98$$

5. As $.98 < 2.71$, we cannot reject the hypothesis.

P R O B L E M S

7.2.1 A die was tossed 42 times and the observed outcomes are listed in the table. Test the hypothesis that the die is fair using a .05 level of significance.

Outcome	Frequency
1	4
2	11
3	8
4	2
5	8
6	9

7.2.2 In order to test if color has any effect on preference, Richard set up the following experiment. Each of four identical boxes was painted a different color: red, blue, green, and black. The subject was told that he would get a prize if he picked the right box. Test the hypothesis, at $\alpha \leq .05$, that color did not make any difference if 100 subjects picked red 32 times, blue 21, green 20, and black 27. (You might give this a real test.)

7.2.3 Suppose that a sample of 200 people are asked if they are for or against abortion. Use the same proportion for "for" and "against" votes as cast in our earlier example. Can we still assume that the community is evenly split at the .05 level of significance? How do you explain this result versus the other result?

7.2.4 A random sample of taxpayers in a school district was asked "Should the school purchase new band oufits for the members of the band?" The results were: agree: 18, disagree: 36, no opinion: 18. What would you choose as a null hypothesis? If you test your hypothesis using χ^2, is the determined χ^2 significant at the .05 level? What conclusions would you draw? Compare your findings with others in your class; did they choose the same hypothesis that you did?

7.2.5 In an elementary school reader, males make decisions 24 times and females 6 times. Is the decision distribution attributable to chance or to other factors? (Remember what Yates would say.)

7.3 Chi-Square Independence Test

We have used χ^2 to determine whether a set of observed outcomes was statistically close to what we would expect to occur. Another use of χ^2 is to determine whether two sets of outcomes are independent of or dependent on each other. The assumptions required in the goodness-of-fit test are also required for the independence test. If data have been collected on how Democrats and Republicans voted on a particular ruling, call it rule A, the chi-square test of independence would provide a means to claim whether the vote on rule A was or was not independent of party affiliation. The vote on rule A might be independent of party affiliation, if both the Democrats and Republicans voted in the same direction, while the vote on rule A might not be independent of party affiliation, if the two parties voted in the opposite direction. The reason we used "might" in the last sentence is that in order to come to a definite conclusion, we need to make use of the chi-square test for independence.

Consider the following situations. A random sample of $n = 100$ community residents has voted for either the Democratic or Republican candidate for mayor, and also voted for or against rule A. The vote is taken and is presented in Table 7.3.1. The question we will answer is: Is the vote on rule A independent of party affiliation? Now that we have the observed frequencies, we must find the expected frequencies. The proportion of the random sample who voted Democratic is $\frac{60}{100}$. *If* party affiliation and choice of vote were *independent*, then $\frac{60}{100}$ of the "yes" vote should be from Democrats and $\frac{40}{100}$ of the "yes" vote should be from Republicans. That is, if voting were independent, then $\dfrac{60(65)}{100} = 39$ Democrats should have voted "yes" and $\dfrac{40(65)}{100} = 26$ Republicans should have voted "yes." Additionally, $\dfrac{60(35)}{100} = 21$ Democrats should have voted "no" and $\dfrac{40(35)}{100} = 14$ Republicans should have voted "no." The expected frequencies are written in parentheses next to the corresponding observed frequencies in Table 7.3.2. The values written alongside this two-row by two-column table are the row sums and the column sums. They are referred to as the *marginal frequencies* (because they are written in the "margin" of the table). If the row sum does not equal the column sum, an error has been made. The degrees of freedom of a two-variable table is found by taking the product of the degrees of freedom in the rows (number of rows − 1) and the degrees of freedom in the columns (number of columns -1). In this problem, there are two row categories (Democrats and Republicans), and there are two column categories (Yes and No), so df $= (r-1)(c-1) = (2-1)(2-1) = 1$.

Table 7.3.1. Rule A

	Yes	No	Total
Democrats	40	20	60
Republicans	25	15	40
Total	65	35	100

Table 7.3.2. Rule A

	Yes	No	Total
Democrats	40 (39)	20 (21)	60
Republicans	25 (26)	15 (14)	40
Total	65	35	100

To determine the value of x^2, the same procedure is used as was used earlier in the goodness-of-fit test. However, in the present situation, the expected values were created by assuming independence. By rejecting the hypothesis (i.e., by obtaining a large value for x^2), we are rejecting the assumption of independence. The independence hypothesis test follows the same procedure that we have been using;

1. H_0: Party affiliation and vote preference are independent.

   H_A: Party affiliation and vote preference are not independent.
2. $\alpha \le .05$; df $= (2-1)(2-1) = 1$.
3. Reject H_0 if $x^2 > 3.84$.
4. Using Yates' correction formula as df $= 1$:

$$x^2 = \frac{(|40-39|-.5)^2}{39} + \frac{(|20-21|-.5)^2}{21} + \frac{(|25-26|-.5)^2}{26}$$

$$+ \frac{(|15-14|-.5)^2}{14}$$

$$= \frac{(.5)^2}{39} + \frac{(.5)^2}{21} + \frac{(.5)^2}{26} + \frac{(.5)^2}{14}$$

$$= \frac{.25}{39} + \frac{.25}{21} + \frac{.25}{26} + \frac{.25}{14} = .046$$

Instead of adding these fractions, notice that each fraction has a value much less than 1. So the sum must be less than 3.84.

5. The assumption of independence cannot be rejected.

Consider the following use of χ^2 to determine if age is independent of favored television shows. Barry, a communications student, asked 500 people which type of television show they liked best. The observed and expected (in parentheses) frequencies are listed in Table 7.3.3. Under the assumption of independence, the expected value of 60* in the second row (13–30 age) and the first column (Westerns) is found as $\frac{200}{500} \cdot 150 = 60$. The "200" is the number people aged 13–30, and since 500 people were sampled, the proportion of people aged 13–30 to the entire group is $\frac{200}{500}$. Since 150 people favored Westerns, assuming independence, $\frac{200}{500} \cdot 150 = 60$ people aged 13–30 should have picked Westerns.

Table 7.3.3. Type of Television Show Liked Best

Age	Western	Drama	Comedy	Variety	Total
Under 13	36 (30)	16 (20)	14 (10)	34 (40)	100
13–30	64 (60)*	34 (40)	20 (20)	82 (80)	200
Over 30	50 (60)	50 (40)	16 (20)	84 (80)	200
Total	150	100	50	200	500

1. H_0: Age and television preference are independent.

   H_A: Age and television preference are not independent.

2. $\alpha \leq .05$; df $= (4-1)(3-1) = 6$.

3. Reject H_0 if $\chi^2 > 12.6$.

4. $\chi^2 = \dfrac{(36-30)^2}{30} + \dfrac{(16-20)^2}{20} + \cdots + \dfrac{(84-80)^2}{80} = 13.3$.

5. Since $13.3 > 12.6$, a value as large as 13.3 would occur by chance at most 5 per cent of the time, and so the independence hypothesis is rejected.

Sometimes the χ^2 independence test is used incorrectly. Consider the following situation. Suppose that 50 females are randomly chosen for a physical education program. Of these 50 females, 30 sleep at least 8 hours per night, while the other 20 females sleep less than 8 hours per night. All 50 women take part in a high-jump exercise. If the number of females who clear the high bar at particular heights is recorded (Table 7.3.4), can we use the test to determine if amount of sleep is independent of jumping ability?

The χ^2 test for independence *cannot* be used in this situation. The numbers of females from column to column are based on the same subjects. That is, since only seven females cleared the 5-foot height who slept less than 8 hours, there could not be more than seven females who cleared

Table 7.3.4. Best Height Cleared

Length of sleep period	Height (ft)			Total
	4	5	6	
8 hr or less	8	7	5	20
At least 8 hr	16	8	6	30
Total	24	15	11	50

the 6-foot height and slept less than 8 hours. In order to use the χ^2 test for independence, it is required that each outcome be independent of each other. This requirement is clearly not met by this particular situation.

PROBLEMS

7.3.1 Are age and television preference independent at the .01 level?

7.3.2 Three groups were trained in developing their ESP. Their responses on an ESP experiment were graded and are reported in Table 7.3.5. Test at both .05 and .01 levels of significance if training was independent of response.

Table 7.3.5

Response	Group		
	1	2	3
Correct	12	6	9
Incorrect	4	9	14

7.3.3 At the end of freshman year at college, 180 of a sample of 300 randomly chosen freshmen stated they date at least four times a month, while the other 120 date less than four times a month. Of those who date at least four times a month, 18 failed a course, while of those who date less than four times a month, 6 failed a course. Test the hypothesis that failing and dating are independent at a .05 level of significance.

7.3.4 Consider the following situation, where a random sample of 100 women is selected from each of the four undergraduate classes at State U. Each woman is asked to express her preference for males who are blond-haired, red-haired, or brown-haired. Can we state that class is independent of preference at the .05 level? Table 7.3.6 contains the observed data. (Before calculating x^2, take a good look at the individual fraction values.)

Table 7.3.6

Class	Preference		
	Blond	Red	Brown
Freshman	30	50	20
Sophomore	60	30	10
Junior	40	30	30
Senior	70	10	20

7.3.5 Construct a situation where the x^2 test for independence cannot be used.

7.4 Sign Test

If a consumer group wanted to test a claim that when a special oil is added to a car, the car will get more miles per gallon than if regular oil is used, they could use the *sign test*. They could randomly choose 12 matched pairs of cars (e.g., two new Vegas, two 1-year-old Datsuns, two 4-year-old Citroens, etc.) and put the special oil in one of each pair and the regular oil in the other of each pair, and then compare which member of each pair got better mileage. Notice that one population is being made into two populations; one population has been randomly chosen to contain the special oil, and the other population has been randomly chosen to contain the regular oil. The sign test compares the effects of two treatments on two groups where members in one group have been matched to members of a second group. The assumption required in order to use the sign test is that the data are continuously distributed. The sign test does not make use of the numerical difference, rather whether the difference between matched pairs is "+" or "−." The data in Table 7.4.1 list the miles per gallon under the two treatments, and the signs of the differences of the matched pairs. The total differences are 7(+), 4(−), and 1(0). If there was, in fact, no difference between the two brands of oil, then the chance that there would be a difference for each matched pair of either + or − would be .50. That is, if the null hypothesis that there was no difference in performance was true, we would expect half the differences to be positive and half to be negative.

We shall test $H_0 : P = .50$ against $H_A : P > .50$ (special oil is better). Since we are interested in the signs of difference, we agree not to count the pairs where the difference was zero. So $n = 11$ instead of 12, as one of the pair

Table 7.4.1

| | Miles per gallon | | Sign of difference |
Matched pair	Special (A)	Regular (B)	($A - B$)
1	18	16	+
2	15	14	+
3	20	20	0
4	19	20	−
5	10	7	+
6	12	13	−
7	15	16	−
8	17	15	+
9	19	16	+
10	8	6	+
11	23	22	+
12	21	23	−

differences was zero. The direction of the alternative hypothesis ($P > .50$) tells us that we shall reject the null hypothesis ($P = .50$) if the number of differences which are positive is very large. How large a difference is the question. As there are only two types of outcomes, + or −, and we are assuming that $P = .50$, we can use the binomial probability table to determine the individual chance of obtaining 0 differences, 1 difference, . . . , 11 differences. We will set $\alpha \leq .05$ and test the hypothesis in our usual manner:

1. $H_0 : P = .50$. (There is no difference between the oils.)
   $H_A : P > .50$. (There is a difference between the oils: special is better.)
2. $\alpha \leq .05$; $n = 11$.
3. Reject the null hypothesis if the number of +'s is at least 9. The critical region is $\{9, 10, 11\}$, since $P(D = 9) + P(D = 10) + P(D = 11) = .027 + .005 + .000 = .032 = \alpha$.
4. The sample data showed that $D = 7$.
5. Do not reject the hypothesis that there is no difference, since 7 does not fall in the critical region.

The sign test is used when the data are continuous and when we want to compare two populations (e.g., special versus usual, A against B) and we have matched pairs to compare. The null hypothesis is always that there is no difference ($P = .50$). However, the alternative hypothesis can be one-tailed or two-tailed. For example, if the sign test involved comparing the effects of a new drug against the effects of an old drug, the new drug could be better ($P > .50$) or it could be worse ($P < .50$). In this case we would use a two-tailed test. The next example makes use of the sign test for a two-tailed test.

Suppose that students in their second year of study are matched on IQ. From these student-matched pairs a random sample of $n = 15$ pairs is separated so that one group is taught by method 1 and the other group is taught by method 2. At the end of the semester, both groups are given the same final examination. The 15 pairs of final grade scores are given in Table 7.4.2.

Table 7.4.2

Matched pair	Method 1	Method 2	Sign of difference
1	74	77	−
2	83	82	+
3	94	93	+
4	68	72	−
5	76	75	+
6	82	80	+
7	79	80	−
8	84	83	+
9	84	84	0
10	96	92	+
11	65	67	−
12	70	71	−
13	83	81	+
14	85	82	+
15	77	75	+

1. $H_0 : P = .50$. (There is no difference between methods.)
   $H_A : P < .50$ or $P > .50$. (One method is better than the other.)
2. $\alpha \leq .10$; $n = 14$.
3. Reject the null hypothesis if the number of +'s is at most 3 or at least 11. The critical region is $\{0, 1, 2, 3, 11, 12, 13, 14\}$, since $P(D = 0) + P(D = 1) + P(D = 2) + P(D = 3) + P(D = 11) + \cdots + P(D = 14) = .058 < .10$.
4. The sample data showed that $D = 9$.
5. Do not reject the hypothesis that there is no difference in the methods since 9 does not fall in the critical region.

While you read this next consideration of the sign test, see if you can think of other matches, aside from amount of sleep, that should be used to test the breakfast diet. Two groups of students were matched on the amount of sleep they get. They were randomly separated into two groups: group A was given a special breakfast diet and group B ate what they usually would (Table 7.4.3). After a two-week period, both groups took an "energy" test. We shall test the hypothesis that the special breakfast diet had no special effect on the energy test as compared to the regular breakfasts.

Table 7.4.3

Matched pair	Special	Regular	Difference
1	51	48	+
2	39	32	+
3	40	41	−
4	38	36	+
5	64	50	+
6	47	39	+
7	62	38	+
8	43	51	−
9	50	48	+
10	70	37	+
11	63	58	+
12	51	54	−

The test follows:
1. $H_0: P = .50$. (There is no difference in energy level.)
   $H_A: P > .50$. (The special diet is more effective.)
2. $\alpha \leq .05$; $n = 12$.
3. Reject H_0 if the number of +'s is at least 10. The critical region is $\{10, 11, 12\}$, since $P(D = 10) + P(D = 11) + P(D = 12) = .016 + .003 + .000 = .019 < .05$.
4. The sample data show that $D = 10$.
5. Reject the hypothesis that the breakfast special diet had no effect. (What other variables might you be interested in knowing about each of these groups—before you run out and buy the new diet?)

PROBLEMS

7.4.1 The following data show the results of test scores of two matched groups of students. Group I did not do homework, while group II did do homework. Test the hypothesis that there is no difference in their test scores at a .05 level of significance.

Matched pair	No homework	Homework
1	72	76
2	75	74
3	73	74
4	76	80
5	84	82
6	81	80
7	72	75

Matched pair	No homework	Homework
8	78	80
9	77	78
10	75	75
11	69	75
12	74	72
13	90	91
14	80	78
15	79	82
16	71	73
17	87	86
18	81	82
19	70	69
20	75	75

7.4.2 The sign test comparing the energy of students who had a special breakfast versus students who had their regular breakfasts is to be retested. This time, instead of matching on amount of sleep, the two groups are matched on the energy test *before* being put on the special versus regular diet. The results of the the energy test after a 2-week period is given below. Retest the hypothesis that the special diet was more effective.

Matched pair	Special	Regular
1	51	48
2	38	36
3	42	43
4	63	60
5	70	68
6	48	48
7	43	42
8	60	59
9	58	59
10	49	46
11	47	45
12	39	37
13	27	28
14	54	53
15	51	51
16	46	44

7.5 Mann–Whitney U-Test

The U-*test* is a stronger statistical test than the sign test. It makes use of the size of the scores, unlike the sign test, which only considers which of the pairs is larger. The U-test provides a way to test if the two sets of scores came from the same population. For example, notice how much closer the matched pair scores are when the diet groups were matched on previous energy levels as compared to the groups matched on amount of sleep. Does this seem reasonable? Why? In using the U-test, the samples that are to be compared need not have the same number of scores, although they might. We can use the U-test to compare two independent samples; matched pairs are not necessary. This test compares two independent samples that have been given different treatments to determine if the two samples are different due to the treatment. The rationale is that if there is, in fact, no (statistical) difference between the two groups due to the treatment, then there will be no (statistical) difference in the ranking of the scores of the two groups. The U-test requires that the data be continuously distributed, as does the sign test.

This test requires that all the scores be ranked with the smallest score being assigned a rank of 1, the next smallest rank of 2, and so on. For example, suppose the following set of scores is found: 2, 10, 7, 8, 6, 6, 5, 3—now we put them in ascending order as we did to determine the median: 2, 3, 5, 6, 6, 7, 8, 10. When we rank them, we give the score of 2 the rank of 1, the score of 3 the rank of 2, the score of 5 the rank of 3. We would not give the first score of 6 the rank of 4 and the second score of 6 the rank of 5 simply because they both have the same value. We average the ranks $\left(\dfrac{4+5}{2} = 4.5\right)$ and give each 6 the rank 4.5. Then the score of 7 is given a rank of 6; and so on. Table 7.5.1 gives the data and the corresponding ranking.

Table 7.5.1

Raw data	Rank
2	1
10	8
7	6
8	7
6	4.5
6	4.5
5	3
3	2

The relationship of U and different rankings are examined in the next two situations. Suppose that groups A and B had scores which are ranked in parentheses as in Table 7.5.2. It is clear from the ranking that group B

Table 7.5.2

A	B
7 (1)	13 (5)
10 (3)	17 (8)
12 (4)	15 (7)
9 (2)	14 (6)
	19 (9)

had a higher rank for each of the scores. We determine the value of U by choosing the smaller of U_1 or U_2, where U_1 is the U-score for group A and U_2 is the U score for group B. The formulas are

$$U_1 = N_1 N_2 + \frac{N_1(N_1+1)}{2} - R_1$$

$$U_2 = N_1 N_2 + \frac{N_2(N_2+1)}{2} - R_2$$

where N_1 is the number of individuals in group A, N_2 the number of individuals in group B, and R_1 is the sum of the ranks of group A, and R_2 is the sum of the ranks of group B. In our situation, $N_1 = 4$, $N_2 = 5$, $R_1 = 10$, and $R_2 = 35$. We find that

$$U_1 = 4 \times 5 + \frac{4(5)}{2} - 10 = 20$$

$$U_2 = 4 \times 5 + \frac{5(6)}{2} - 35 = 0$$

where U is the smaller of U_1 and U_2, and thus $U = U_2 = 0$. So when the ranking is distributed so that the one group has all the low ranks and the other group has all the high ranks, the value of U appears to be zero. Suppose the scores are changed so that neither group is consistently higher in rank (Table 7.5.3). In this situation $N_1 = N_2 = 4$, but $R_1 = 18$ and $R_2 = 18$, such that

$$U_1 = 4 \times 4 + \frac{4(5)}{2} - 18 = 8$$

$$U_2 = 4 \times 4 + \frac{4(5)}{2} - 18 = 8$$

Table 7.5.3

A	B
7 (1)	9 (2)
12 (4)	10 (3)
13 (5)	14 (6)
17 (8)	15 (7)

In this case $U = U_1 = U_2 = 8$—neither group has a ranking pattern that is different from the other. The closer U is to zero, the more likely there is a real difference between the groups.

Suppose that following data represent the hours that two groups of students could go without sleep. One group practiced transcendental

Table 7.5.4. Hours
Without Sleep

T.M.	Without T.M.
27 (8)	16 (1)
18 (2)	35 (10)
41 (11)	28 (9)
20 (4)	22 (6)
21 (5)	26 (7)
19 (3)	

mediation (T.M.) and the other group did not (Table 7.5.4). In this situation, $N_1 = 6$, $N_2 = 5$, $R_1 = 33$, and $R_2 = 33$. We obtain that

$$U_1 = 6 \times 5 + \frac{6 \times 7}{2} - 33 = 18$$

$$U_2 = 6 \times 5 + \frac{5 \times 6}{2} - 33 = 12$$

The value of U that we use for the U-test is the smaller of U_1 and U_2, and so in this problem we take $U = 12$. Table VI in Appendix B has n representing the size of the smaller group and m representing the size of the larger group. So, in our problem we have $m = 6$ and $n = 5$. For $\alpha \le .05$ and a two-tailed test, we find that the intersection where $m = 6$ and $n = 5$ is $U = 3$. In order for there to be a significant difference between the two groups at the .05 level, the calculated U must be smaller than the table U. Since our $U = 12$ and $12 \not< 3$, we cannot reject the hypothesis that there is no significant difference between the two groups. Notice that our data in

Table 7.5.2 produced a $U = 0$ when $m = 5$, $n = 4$, and since the calculated U (0) is less than the table U (1), there was a significant difference in the ranking of the groups.

The following problem using the U-test will be examined using our usual form for hypothesis testing. Two groups of people are being compared as to which group is best at holding breath. The first group is on a special training program, while the second group is not doing any special training. The time in seconds, and rank, of the members of the two groups is listed in Table 7.5.5. We shall test the hypothesis that there is no difference between the groups. The test will be one-tailed to the right, as the alternative hypothesis is that the special training group can hold their breaths significantly ($\alpha \leq .05$) longer.

Table 7.5.5. Time Holding Breath

Special training	No training
47 (4)	36 (3)
67 (9)	28 (2)
59 (7)	60 (8)
83 (10)	48 (5)
50 (6)	16 (1)

1. H_0: No difference in rank.

   H_A: Special training group is significantly higher in rank.
2. $\alpha \leq .05$; $N_1 = 5$; $N_2 = 5$.
3. $U_1 = 5 \times 5 + \dfrac{5 \times 6}{2} - 36 = 4$; $U_2 = 5 \times 5 + \dfrac{5 \times 6}{2} - 19 = 21$.
4. Reject the null hypothesis if $U > 4$.
5. Since $U_1 < U_2$, we take $U = U_1 = 4$. And as 4 is not less than 4, the hypothesis cannot be rejected at the .05 level of significance. (Did you think it was going to be rejected by just looking at the rankings?)

PROBLEMS

7.5.1 Two groups are being wakened each morning by different methods. Group 1 is wakened by a siren and group 2 by soft music. The times it takes for the individuals in each group to get out of bed are listed in Table 7.5.6. Determine if there is a significant difference in time needed to get out of bed by the U-test for a two-tailed test at $\alpha \leq .05$ and at .10.

7.5.2 Use the sign and U-tests to determine if there is a significant difference between the two groups in Table 7.5.7 at a .05 level of significance for a two-tailed test.

Table 7.5.6. Time (seconds)
to Get Out of Bed

Siren	Soft music
3	6
4	10
8	16
9	11
1	15
2	136

Table 7.5.7

Group I	Group II
10	12
5	7
3	2
14	11
9	13
8	1
6	4
16	15

7.6 Summary

In this chapter we have examined some distribution-free tests. Distribution-free tests are tests which do not require that the sample data come from a particular type of distribution. The four hypothesis tests we considered were the chi-square test for goodness of fit, the chi-square test for independence, the sign test, and the U-test. Each type of test allowed us to test hypotheses under special circumstances. In all the tests we must assume that we have a random sample from the appropriate population.

The chi-square test for goodness of fit provided a way of comparing whether or not the observed outcomes were significantly different from the expected outcomes. The chi-square test for independence tested whether one variable was independent or dependent upon another. The sign test examined whether or not different treatments cause significant differences between two matched-pair groups. A unique aspect of this test is that differences due to treatments between matched pairs are measured in greater-or-less-than terms rather than in terms of numerical differences, as with most of the tests that we have considered. The U-test provided a hypothesis test for comparing two groups taken from independent samples.

This test considered the magnitude of the scores for the two groups under different treatments. In this test it was not necessary for the number of each group to be the same, as was the case for the sign test.

7.7 Review of Formulas

I. Chi-square test of goodness of fit
   A. Assumptions
      1. Outcomes are independent of each other.
      2. Every outcome falls into one and only one category.
      3. The frequency of every expected category should be at least 5.
   B. Calculation of chi-square

      1. $\chi^2 = \dfrac{(o_1 - e_1)^2}{e_1} + \dfrac{(o_2 - e_2)^2}{e_2} + \cdots + \dfrac{(o_n - e_n)^2}{e_n} = \sum\limits_{i=1}^{n} \dfrac{(o_i - e_i)^2}{e_i},$

      where there are n categories, o_i is the frequency of the observed outcomes for the ith category, and e_i is the frequency of the expected outcomes for the ith category. The degrees of freedom, df $= n - 1$. The sum of the observed frequency must equal the sum of the expected frequency.

      2. If the calculated χ^2 was greater than the χ^2 determined by α, the level of significance, the hypothesis was rejected. That is, a significantly large χ^2 tells us that there was not a good fit between what we observed and what we expected.
   C. Yates' correction formula
      If the degrees of freedom $= 1$, then in place of the chi-square formula (B1), we would use

      $$\chi^2 = \frac{(|o_1 - e_1| - .5)^2}{e_1} + \cdots + \frac{(|o_n - e_n| - .5)^2}{e_n} = \sum_{i=1}^{n} \frac{(|o_i - e_i| - .5)^2}{e_i}$$

II. Chi-square test of independence
   A. Assumptions are the same as stated in IA.
   B. Calculation of chi-square

      1. $\chi^2 = \sum\limits_{i=1}^{r} \sum\limits_{j=1}^{c} \dfrac{(o_{ij} - e_{ij})^2}{e_{ij}}$, where e_{ij} is the expected frequency in the ith row and jth column. The df $= (r - 1)(c - 1)$.

      2. If the calculated χ^2 was greater than the χ^2 determined by α and the degrees of freedom, then the hypothesis claiming independence was rejected.

III. Sign test
   A. Assumptions
      1. Data are continuously distributed.
      2. Groups contain matched pairs.

B. Sign-test rationale

If the number of differences (+ or −) between the matched pairs in the two groups fell in the critical region determined by H_A, the alternative hypothesis; α, the level of significance; and n, the sample size, the hypothesis that stated there was no difference between methods was rejected. The null hypothesis was always set at $P = .50$, and the binomial probability tables were used to determine the critical region.

IV. U-test

A. Assumptions

1. Data are continuously distributed.
2. Samples are independent.

B. Calculation of U

1. $U =$ smaller of U_1 or U_2, where

$$U_1 = N_1 N_2 + \frac{N_1(N_1 + 1)}{2} - R_1$$

$$U_2 = N_1 N_2 + \frac{N_2(N_2 + 1)}{2} - R_2$$

and N_1 is the number of elements in group 1, N_2 the number of elements in group 2, R_1 the sum of the ranks of group 1, and R_2 the sum of the ranks of group 2.

2. If the calculated U was less than the table U determined by N_1, N_2, and α, the hypothesis claiming that there was no difference between groups was rejected.

7.8 Problem Set

7.8.1 The director of a hospital wants to determine if babies are born in the same proportion in the three 8-hour shifts as they used to be. If the proportions have been changed significantly, she will redistribute the staff. The usual proportion of children born during the three shifts has been .25, .40, and .35. In a recent week, picked at random, the three shifts accounted for 80, 114, and 106 births, respectively. Is this week different from the usual response at the .10 level of significance?

7.8.2 If the hypothesis has been rejected at the .05 level, can you make a different decision of an $\alpha \le .10$ test? Why? What if the α values are reversed?

7.8.3 Ten years ago the ratio of students who were reading on levels 1, 2, 3, 4, and 5 were $1:4:6:4:1$, respectively. A sample of 160 students taken recently produced results of 13, 45, 63, 34, and 6 for the respective reading levels. Can we be 95 per cent confident that the earlier ratio still holds?

7.8.4 A departmental examination was given to classes A and B. Test the hypothesis that grade on the examination was independent of class for $\alpha \leq .05$.

Class	Passed	Failed
A	72	28
B	58	22

7.8.5 In rating a particular organization "favorable" or "unfavorable," is education independent of how the organization is rated? ($\alpha \leq .05$.)

	Grade school	High school	College
Favorable	28	24	14
Unfavorable	72	76	86

7.8.6 A group of psychology students were asked to respond to a multiple-choice question. The interviewer was interested in determining if sex was independent of choice. Determine if independence holds for $\alpha \leq .05$ (a) by grouping and (b) by Yates' correction formula.

	Multiple-choice response				
Sex	1	2	3	4	5
Male	12	33	17	8	0
Female	42	75	16	4	3

7.8.7 A test for aggressiveness is given to 10 matched pairs of brothers and sisters. Test the null hypothesis that there is no difference between the two groups at $\alpha \leq .10$ for (a) a one-tailed test (which direction do you think H_A should be?) and (b) a two-tailed test.

Pair	Male	Female
1	37	32
2	43	40
3	29	31
4	36	35

Pair	Male	Female
5	52	39
6	40	40
7	47	48
8	57	49
9	51	50
10	33	29

7.8.8 A teacher randomly divides a class so that part of the class studies in a group, and the others study independently. At the end of 2 weeks, he gives the class a test. Determine at a .05 level of significance if there was a significant difference in learning as measured by the test scores.

Group study	Independent study
81	70
79	69
82	96
65	83
93	75
74	

7.8.9 Ten persons got together and agreed to give up smoking. They had heard that some persons gain weight when they stop smoking. The weights of the individuals at the time they decided to stop smoking and 2 months after they had stopped are listed in the table. Test at the .05 level of significance that giving up smoking has no effect on a person's weight against the alternative hypothesis that giving up smoking does increase one's weight.

Before	After
152	160
170	170
169	173
182	180
145	134
202	211
194	197
188	187
166	172
203	207

Could you think of other information that you could have collected that might make you feel more confident about your decision?

7.8.10 The president of a company wants to determine whether there is a connection between years on the job and a worker's opinion concerning a proposed change in working schedule. She interviews the 200 employees and obtains the results shown in the following table:

	0–6	6–12	12–18	18–24
Against	4	10	8	6
Indifferent	20	21	12	8
For	16	39	30	26

Use a level of significance of .05 to determine if the null hypothesis that there are no differences of opinion concerning the change among the four groups should be rejected.

7.8.11 Reanalyze the table in the last problem leaving out the indifferent responses.

7.8.12 The following information is the distribution of male cats in 80 litters, each containing 4 cats:

Number of male cats	Number of litters
0	6
1	18
2	28
3	23
4	5

Assuming that the birth of a male cat equals that of a female cat, determine if the expected outcomes agree with the assumption of equality.

7.8.13 If chi-square analysis concludes that there is a dependent relationship between the variables, the strength of the relationship can be measured by the contingency coefficient: $C = \sqrt{\dfrac{\chi^2}{\chi^2 + n}}$, where n is the total frequency. The larger the value of C, the stronger the dependence. Calculate C for those analyses which showed dependence in the problems you considered in 7.3.1 to 7.3.5, in the problem where the type of television show liked best was found to correspond with the age group, and in any of the problems of this section testing for independence that were rejected.

7.8.14 Should C be calculated for the following problem: A random sample of 209 graduates of Loe High School revealed that 30 of the 103 students who were not on athletic teams went to college, while 55 of the 156 who were on athletic teams went to college? At a .05 level of significance is it justified to say that a significantly greater proportion of athletic-team members went to college than students who were not on athletic teams?

7.8.15 What are the lower and upper values that C may take?

7.8.16 Random samples of seventh graders from two different schools scored the following grades on a city-wide achievement test:

School A: 77, 73, 59, 93, 88, 94, 98, 97, 92, 71, 99, 77, 98, 91, 63, 70, 91, 93, 79, 97, 98, 85, 74, 97, and 81

School B: 90, 75, 46, 57, 72, 75, 97, 95, 89, 67, 63, 52, 89, 64, 64, 89, 38, 64, 76, 79, 35, 76, 64, 69, and 77

Randomly match each student from School A with a student from School B, and use the sign test to determine if there is or is not a significant difference between the achievement test scores between the two schools at a level of significance of .05.

7.8.17 A group of students was randomly divided so that Group 1 studied number theory for 4 weeks while Group 2 studied economics theory for 4 weeks. At the end of that time, both groups were asked to pick 20 slips of paper containing numbers from 1 to 100, while blindfolded, and to try to pick those numbers that were prime. The following are their scores of number of primes picked:

Group A: 4, 8, 6, 9, 3, 8, 10, 7, 12
Group B: 5, 1, 6, 2, 8, 6, 9, 8, 10

Use the U-test with a level of significance of .05 to determine if there is not a difference between the two groups. (U-test?)

7.8.18 Group A was taught history with both a text and slides, while Group B was taught with just the text. A test was given to both groups. Determine whether the inclusion of slides led to greater learning for Group A, at a .05 level of significance.

Group A: 77, 86, 43, 72, 95, 79, 93, 71, 54, 90, 61, 69
Group B: 55, 46, 40, 57, 92, 63, 53, 70, 29, 70, 80, 42

Chapter 8
Survey Taking

8.1 Introduction

Taking a survey is a complex and time-consuming activity but it is also an exciting and meaningful learning experience. You have to make many decisions: what population you will sample, how you will choose the sample, what it is important to find out, how to write questions clearly, how to organize the data, and what procedures to use to evaluate the sample data—to name a few. In the process of doing a survey, you can get a good idea of how your community feels about certain issues compared to how you thought they felt, and observe, firsthand, how your knowledge of statistics can be put into practice. The information you find may provide insights into people's feelings on important issues such as equal rights, day care, and national health insurance. An average-sized class could collect the responses of a sample of 250 to 1,000 people, which is large enough to be of real value.

The following are important elements in a good survey:

1. The sample is as unbiased as possible.
2. The questions are well thought out and clearly stated, and there has been given a trial run to ensure clarity.
3. The choice of replies to a question is limited to five at the most.
4. The interviewer presents the questions in a *uniform* manner to the respondents.

This chapter will explore some sampling techniques and make a number of suggestions for taking a survey. We shall begin with the crucial aspect of survey taking: trying to ensure that the survey data are not biased, that is, that the method of choosing who will be in the sample ensures that the sample accurately reflects the community from which the sample is taken. There is a phrase in the computer field that also relates to survey taking: "garbage in, garbage out." In other words, if the data put in the computer are poorly organized, there is no way that the computer can make it valuable. The same goes for survey taking. If the procedure for determining who will be in the sample is not carefully thought out, the results will not be any better.

8.2 Random Sampling

The collection of data may range from highly successful to disastrous. However, success or failure does not so much depend on the physical process of data collection as on the method used to determine who will be in the sample. We mentioned in Chapter 1 the poor prediction made by the *Literary Digest* regarding the election of Franklin Roosevelt. The problem was that the method they used to collect data did not take into consideration the average American in the 1930s and, as would be expected, their results were biased. Their findings represented the opinions of a particular class of people, the wealthy people, not a cross section of the entire population, as they expected.

Compare the sample of approximately 2,000,000 taken by the *Literary Digest* to the sample of approximately 1,500 taken by poll-takers today. The effectiveness of present-day sampling methods is shown by the results obtained by the 1958 Gallup poll of the Congressional election; the difference between their findings and the actual vote was one half of one percentage point. Although we cannot expect to do what the major polling organizations do in choosing their sample, we can make some attempts to guarantee that we do not create a sampling fiasco such as that of the *Literary Digest.*

One of the sampling techniques that you may use which eliminates sampling bias is *random sampling.* The concept of random sampling does not mean "choose anyone from the population." A truly random sample means that every individual in the population has an equal chance of being chosen. A student who gives his survey questionnaire to friends or any student sitting nearby in the cafeteria has not collected a random sample but, rather, a haphazard sample. Once randomness is part of your sampling procedure, sampling bias has been eliminated. Random sampling does not ensure that all groups within the total group will be proportionately represented in the sample, only that those who are selected are chosen on the basis of chance alone, and therefore without bias.

One way to go about collecting a random sample is to use a random-numbers table. In these tables all 10 digits, 0 through 9, are presented so that each digit occurs very close to 10 per cent of the time, and the occurrence of any one digit does not appear to affect the occurrence of any other. The effect of this is that by choosing members of the population using the random-numbers table provides a random-sampling procedure absent of bias. Table VII in Appendix B lists 2,500 random numbers. If the digits are randomly distributed, over the long run each of the 10 digits will occur 10 per cent of the time. For example, although the first 20 digits do not contain a "4," a "4" occurs .082 ($\frac{41}{500}$) in the first 500 digits, .090 ($\frac{90}{1,000}$) in the first 1,000 digits, and .099 ($\frac{124}{1,250}$) in the first 1,250 digits. As you can see, as we take larger samples, the chance of picking a "4" moves toward .10, the expected proportion.

Suppose there are 1,000 students in your school and you wish to take a random sample of 100 of them. You will need a list of all the student names. Then, from the random-numbers table, you would select a row or column randomly (moving vertically or horizontally). One way to randomly choose where to begin is to take a pin, close your eyes, and then stick the pin in the table. If the pin is closest to a "2," begin on a horizontal or vertical line where "2" is the first digit. Then divide the digits into sets of three. We choose three-digit numbers because there are 1,000 three-digit numbers beginning with 000 through 999, and this method would fit if the number of people in the population was at most 1,000. For example, part of the first row of the table beginning with "2" would be written 235 237 831 773 208.... These numbers are used to locate the corresponding name on the class list—the 235th name is the first person picked for the sample; and so on.

If your school population (or whatever population you are examining), for example, contained 1,500 persons, the method suggested would not be suitable, as it does not allow you to choose above the 1,000th student. What you could do in this case is the following: Choose four-digit numbers from 0000 to 9999 and omit any number above 1,500. In this way, each of the 1,500 students will have an equal chance of being picked. Choosing every 10th or 15th student from a school list does not produce a strictly random sample. However, if you do choose this method, make sure that the initial number chosen has been chosen randomly.

Another sampling technique that incorporates random sampling is *proportional sampling*. As the name implies, this method is intended to give proportional representation to each group in the total population. For example, if your school population totals 1,800, where there are 600 freshmen, 540 sophomores, 360 juniors, and 300 seniors, the proportion of your sample should correspondingly include $\frac{1}{3}$ freshmen, .30 sophomores, .20 juniors, and $\frac{1}{6}$ seniors. Thus a sample of 150 students would include 50 freshmen, 45 sophomores, 30 juniors, and 25 seniors. Once you have determined the number in each group, you should use the random-digits table to decide exactly who the sample will contain. While this sampling procedure is more involved than the random-sampling method, its merit lies in the fact that it guarantees proportional representation, whereas the random-sampling procedure guarantees only that there will be no sampling bias.

8.3 Sampling for the Nonprofessional

If your survey is done carefully and objectively, it will be very worthwhile. Although we are not professional survey takers, we feel that you do not have to eat a whole pineapple to know its taste, and we would rather that you will be able to detect a poor one.

Before we suggest some ways of going about collecting data, here is a list of things not to do. Experience has shown that certain situations should be avoided:

1. Do not interview families during mealtime, too late in the evening, or too early on the weekend.

2. Do not ask questions that are too personal. The more personal the questions, the less personal the answer. To give you an example of what we mean, Darrell Huff and Irving Geis, in *How To Lie With Statistics* (New York: W. W. Norton, 1954), mention that in response to the personal question, "How many showers do you take a day?" the average number for the sample turned out to be between two and three!

3. Do not bias the questions by suggesting in the question a certain answer. For example, the question about showers implies that people take more than one shower per day, since the question is stated in terms of "showers" and not "shower."

4. Do not try to help the person make a decision by urging or by "explaining" the choices.

5. Do not create overly narrow ranges for questions dealing with age or income; for example, 21 to 30 would be a better age range than 21 to 23.

6. Do not ask the survey questions of people you know—this may bias them. For, you cannot be sure whether the answers truly represent their personal opinions or whether they are what the person thinks you want to hear.

7. Do not (e.g., when interviewing people in front of a store) ask one person who is part of a group to answer the questionnaire and leave the others out. The person picked out might feel rushed or embarrassed if the others are standing around. You might let them all fill out the questionnaire and only keep the person who was designated by your sampling method.

8. Do not take the opinion of only those people who live in the same vertical line in an apartment house. Apartments in each vertical (e.g., 1C, 2C, 3C, etc.) are usually all of the same size. Therefore, the sizes of the families and their incomes might be much closer to each other than to those from other apartments. If you are taking a sample in a residential area, do not include all the corner houses. (Why?)

9. Do not let a "No opinion" statement confuse you. According to John Fenton, who was involved in the Gallup polls and wrote *In Your Opinion . . .*, there are three types of no opinion—"the uncomprehending, the uninterested, and the undecided." If the person is uninterested, move on; if the person does not understand the question, you have to decide if you want to explain the situation; and, if the person is undecided, you can count his statement as undecided or try repeating the question at the end of the survey.

If you were interested in the opinions of a cross section of the community, a map of the area would be helpful. If possible, choose a proportion of people from different sections which might reflect the differences in population distribution. You might station yourself at a major location in each area (ideally, where people tend to have to wait—e.g., a bus stop or checkout counter) and interview those people who correspond to the set of random numbers you have. Time as well as location is important. If you are doing your survey in the early afternoon, the adult males available might not be representative of the entire population of adult males. Those who are available might be on vacation, out of work, employed in the neighborhood, and so on. Some people have set up patterns where they do certain things at certain times and so you might only see people whose pattern fits yours if you collected your surveys at the time over a period of days. Perhaps some students could collect responses in the morning, some in the afternoon, and others at night on different days so as to get a more complete survey of the community. Whenever you read about a random sample survey being done, remember that the responses are made by people who wanted to respond or did not mind being asked the questions. Those who did mind did not respond. This is the problem with mailing surveys. There is no guarantee of randomness. Only the people who want to respond do respond.

Your questionnaire should contain some information regarding the background of the person being questioned. For example, you might record the person's sex, level of education, political party, and the like. This information provides you with extra data that can help you get further insight into the sample. You could use the distribution-free tests discussed in Chapter 7 to determine if responses are independent of sex, education, and so on, or if there is a difference between how husbands and wives respond to certain questions. As we mentioned earlier, background information such as age should have reasonable ranges. For example, age ranges might be 21 to 30, 31 to 40, etc., and education ranges might be grade school, high school, college, etc. Background information allows you to make statments such as: "Of the people who are 21 to 30 years of age, k per cent believe _____, while of the people who are 51 to 60, j per cent believe _____."

8.4 Statistical Tests

The question of whether you can use the binomial and/or normal distribution requires some consideration. As we mentioned, the distribution-free tests do not require that the population distribution be known. If the data lend themselves to categories, the chi-square tests can be used (if the assumptions are met), whereas if the data can be said to be continuous, the sign and U-tests can be used.

We said that the binomial distribution required certain assumptions about the population to be known before it could be used. They are:

1. There are only two types of outcomes.
2. The number of trials is known.

3. The trial outcomes are independent of each other.

4. The probability of success is constant for each trial.

The first requirement can be satisfied easily. For example, suppose there were four choices available: A, B, C, and D. If you wanted to examine response A using the binomial model, you could put B, C, and D into one category of "not-A." Since the sample size is predetermined, the second requirement is satisfied. As for the third requirement, we can think of the trials as being independent inasmuch as choosing a particular person does not affect your choosing or not choosing another person if you use a random-selection process. The fourth requirement, that the probability of success be constant, requires a comparison of the effects on the probability of an outcome of sampling with and without replacement.

The binomial distribution requires that the chance of success be constant for all trials. The question that must be answered is: Since a person is not picked again after he responds, how can the chance of "success" be constant? We will show that if the sample size is small relative to the population size, the probability of success stays relatively constant, and so we can use the binomial probability model. Consider a population of 1 million voters, where 50 per cent are for candidate A and 50 per cent are for candiate B. If we take a random sample of $n = 500$ voters, can we assume that $P(A) = P(B) = .50$ stays (relatively) constant? Suppose that we replace each person after he or she is picked; this will keep the total at 1 million voters and allow us to pick the person again. This guarantees that $P(A) = .50$. By sampling with replacement, the probability of picking 500 people who agree on candidate A is $\dfrac{500,000}{1,000,000} \times \dfrac{500,000}{1,000,000} \times \cdots \times$

$\dfrac{500,000}{1,000,000}$. However, if we do not use replacement, the sample size shrinks, as does the number of people who are left and are for candidate A. The probability of picking 500 people who agree on candidate A, without replacement, is

$$\frac{500,000}{1,000,000} \times \frac{499,999}{999,999} \times \frac{499,998}{999,998} \times \cdots \times \frac{499,501}{999,501}$$

The greatest difference between sampling with and without replacement is at the 500th vote—$\dfrac{500,000}{1,000,000}$ (replacement) versus $\dfrac{499,501}{999,501}$ (without replacement). But, $\dfrac{500,000}{1,000,000} = .5$ and $\dfrac{499,501}{999,501} = .49975$. You can see that the difference is very slight. As long as the sample size is small relative to the population size, it is acceptable to assume that the probability of success is constant, as if each person were replaced and could be picked again. For the binomial distribution, the sample size should not be 20 per cent or more of the population size. As long as it is less, the probabilities found with and without replacement are close enough to say that the probability of success is constant.

To see the difference when the sample size is large relative to the population size, consider the following situation. Suppose that we have a population of 10 voters, half of whom are for candidate A and the other half for candidate B. We shall compare the probability of picking a sample of $n = 4$, all of whom are for candidate A, first using replacement and then without replacement. By replacement, we have $\frac{5}{10} \times \frac{5}{10} \times \frac{5}{10} \times \frac{5}{10}$ as the probability. Without replacement, we have $\frac{5}{10} \times \frac{4}{9} \times \frac{3}{8} \times \frac{2}{7}$. You can see that the associated pairs move farther apart from each other: $\frac{5}{10}$ and $\frac{5}{10}$ to $\frac{5}{10}(.500)$ and $\frac{2}{7}(.286)$. Unlike the first example, where the sample was small relative to the population, in this problem, without replacing the voters, 40 per cent (4 of 10) of the population is diminished. This large decrease shows up in that by sampling without replacement the chance of choosing the fourth person for candidate A is .286 as versus .500 if there had been sampling with replacement. That is, when the sample is large relative to the population, there is a considerable difference in the probabilities when there is replacement and when there is no replacement. Thus, when taking a survey, it is reasonable to assume that the binomial probability model will be useful whenever the sample size is small compared to the population size.

We shall have to limit the use of the normal distribution in our survey taking. Although we do not know that the population is normally distributed as to the responses on the questionnaire, the normal distribution can be used to provide approximations for the binomial distribution when n is sufficiently large. Confidence intervals provide additional information that hypotheses tests do not. We suggest that in a survey you consider both. It is possible, if the sample size is small, not to reject two hypotheses that may differ by 40 per cent. When the confidence interval is determined and the sample size is small, the interval will be very large, indicating that many very different claims about the population can be made.

Before doing a survey, you might find it interesting to write down what you think the results will be, along with the percentages, and compare your hypotheses with the actual findings.

Appendix A

Note on Summation Notation

From time to time as our development proceeds, we use a special notation to simplify the writing of sums. This notation is standard, but since it is occasionally omitted from the traditional algebra course, we shall include its essential features.

We use $\sum\limits_{j=1}^{n} j$ to represent the sum $1+2+3+4+\cdots+n$; 1 and n are called the *upper and lower limits* of summation, and j is called the *index* of summation. For example, $\sum\limits_{j=1}^{4} j = 1+2+3+4 = 10$; $\sum\limits_{j=3}^{6} j = 3+4+5+6 = 18$; $\sum\limits_{i=0}^{4} a_i = a_0+a_1+a_2+a_3+a_4$ and, analogously, $\sum\limits_{i=1}^{3} i^2 = 1^2+2^2+3^2 = 14$.

It should be clear that we replace the index of summation by consecutive integers, starting with the lower limit of summation and ending with the upper limit, and then we add the resulting expressions. Conversely, we can use summation notation to write $1+2+3+4+5$ as $\sum\limits_{j=1}^{5} j$, and $3^2+5^2+7^2+9^2$ as $\sum\limits_{i=1}^{4} (2i+1)$.

A constant factor that is independent of the index of summation can be factored out of the entire summation. For example,

$$\sum\limits_{j=1}^{5} 2j = 2(1)+2(2)+2(3)+2(4)+2(5) = 2(1+2+3+4+5) = 2\sum\limits_{j=1}^{5} j$$

and, generally,

$$\sum\limits_{i=1}^{m} ka_i = k\sum\limits_{i=1}^{n} a_i = k(a_1+a_2+a_3+\cdots+a_n)$$

Special meaning is given to the summation of a constant, which we can look upon as a constant factor independent of the index of summation. In this case the result is equal to the value of the constant times the number of values the index assumes. For example, $\sum\limits_{i=1}^{3} 5 = 5+5+5 = 15$, $\sum\limits_{i=4}^{7} 3 = 3 \cdot 4 = 12$, and $\sum\limits_{j=0}^{3} 5x = 20x$.

Another important summation formula is

$$\sum_{i=1}^{n} (a_i + b_i) = (a_1 + b_1) + (a_2 + b_2) + \cdots + (a_n + b_n)$$

$$= (a_1 + a_2 + \cdots + a_n) + (b_1 + b_2 + \cdots + b_n)$$

$$= \sum_{i=1}^{n} a_i + \sum_{i=1}^{n} b_i$$

For example,

$$\sum_{i=1}^{4} (2i + 3) = \sum_{i=1}^{4} 2i + \sum_{i=1}^{4} 3 = 2 \sum_{i=1}^{4} i + \sum_{i=1}^{4} 3 = 2(1+2+3+4) + 3 \cdot 4 = 32$$

and

$$\sum_{j=1}^{3} (3j^2 - 1) = \sum_{j=1}^{3} 3j^2 + \sum_{j=1}^{3} (-1) = 3(1^2 + 2^2 + 3^2) + (-1)(3) = 39$$

Appendix B
Tables

Table I. Squares and Square Roots

n	n^2	$\sqrt{n}$	$\sqrt{10n}$	n	n^2	$\sqrt{n}$	$\sqrt{10n}$
1.0	1.00	1.000	3.162	5.5	30.25	2.345	7.416
1.1	1.21	1.049	3.337	5.6	31.36	2.366	7.483
1.2	1.44	1.095	3.464	5.7	32.49	2.387	7.550
1.3	1.69	1.140	3.606	5.8	33.64	2.408	7.616
1.4	1.96	1.183	3.742	5.9	34.81	2.492	7.681
1.5	2.25	1.225	3.873	6.0	36.00	2.449	7.746
1.6	2.56	1.265	4.000	6.1	37.21	2.470	7.810
1.7	2.89	1.304	4.123	6.2	38.44	2.490	7.874
1.8	3.24	1.342	4.243	6.3	39.69	2.510	7.937
1.9	3.61	1.378	4.359	6.4	40.96	2.530	8.000
2.0	4.00	1.414	4.472	6.5	42.25	2.550	8.602
2.1	4.41	1.449	4.583	6.6	43.56	2.569	8.124
2.2	4.84	1.483	4.690	6.7	44.89	2.588	8.185
2.3	5.29	1.517	4.796	6.8	46.24	2.608	8.246
2.4	5.76	1.549	4.899	6.9	47.61	2.627	8.307
2.5	6.25	1.581	5.000	7.0	49.00	2.646	8.367
2.6	6.76	1.612	5.099	7.1	50.41	2.665	8.426
2.7	7.29	1.643	5.196	7.2	51.84	2.683	8.485
2.8	7.48	1.673	5.292	7.3	53.29	2.702	8.544
2.9	8.41	1.703	5.385	7.4	54.76	2.720	8.602
3.0	9.00	1.732	5.477	7.5	56.25	2.739	8.660
3.1	9.61	1.761	5.568	7.6	57.76	2.757	8.718
3.2	10.24	1.789	5.657	7.7	59.29	2.775	8.775
3.3	10.89	1.817	5.745	7.8	60.84	2.793	8.832
3.4	11.56	1.844	5.831	7.9	62.41	2.811	8.888
3.5	12.25	1.871	5.916	8.0	64.00	2.828	8.944
3.6	12.96	1.897	6.000	8.1	65.61	2.846	9.000
3.7	13.69	1.924	6.083	8.2	67.24	2.864	9.055
3.8	14.44	1.949	6.164	8.3	68.89	2.881	9.110
3.9	15.21	1.975	6.245	8.4	70.56	2.898	9.165
4.0	16.00	2.000	6.325	8.5	72.25	2.915	9.220
4.1	16.81	2.025	6.403	8.6	73.96	2.933	9.274
4.2	17.64	2.049	6.481	8.7	75.69	2.950	9.327
4.3	18.49	2.074	6.557	8.8	77.44	2.966	9.381
4.4	19.36	2.098	6.633	8.9	79.21	2.983	9.434
4.5	20.25	2.121	6.708	9.0	81.00	3.000	9.487
4.6	21.16	2.145	6.782	9.1	82.81	3.017	9.539
4.7	22.09	2.168	6.856	9.2	84.64	3.033	0.592
4.9	23.04	2.191	6.928	9.3	86.49	3.050	9.644
4.9	24.01	2.214	7.000	9.4	88.36	3.066	9.695
5.0	25.00	2.236	7.071	9.5	90.25	3.082	9.747
5.1	26.01	2.258	7.141	9.6	92.16	3.098	9.798
5.2	27.04	2.280	7.211	9.7	94.09	3.114	9.849
5.3	28.09	2.302	7.280	9.8	96.04	3.130	9.899
5.4	29.16	2.324	7.348	9.9	98.01	3.146	9.950

Table II. Binomial Probability Values

n	x	.05	.10	.20	.30	.40	*p* .50	.60	.70	.80	.90	.95
1	0	.9500	.9000	.8000	.7000	.6000	.5000	.4000	.3000	.2000	.1000	.0500
	1	.0500	.1000	.2000	.3000	.4000	.5000	.6000	.7000	.8000	.9000	.9500
2	0	.9025	.8100	.6400	.4900	.3600	.2500	.1600	.0900	.0400	.0100	.0025
	1	.0950	.1800	.3200	.4200	.4800	.5000	.4800	.4200	.3200	.1800	.0950
	2	.0025	.0100	.0400	.0900	.1600	.2500	.3600	.4900	.6400	.8100	.9025
3	0	.8574	.7290	.5120	.3430	.2160	.1250	.0640	.0270	.0080	.0010	.0001
	1	.1354	.2430	.3840	.4410	.4320	.3750	.2880	.1890	.0960	.0270	.0071
	2	.0071	.0270	.0960	.1890	.2880	.3750	.4320	.4410	.3840	.2430	.1354
	3	.0001	.0010	.0080	.0270	.0640	.1250	.2160	.3430	.5120	.7290	.8574
4	0	.8145	.6561	.4096	.2401	.1296	.0625	.0256	.0081	.0016	.0001	
	1	.1715	.2916	.4096	.4116	.3456	.2500	.1536	.0756	.0256	.0036	.0005
	2	.0135	.0486	.1536	.2646	.3456	.3750	.3456	.2646	.1536	.0486	.0135
	3	.0005	.0036	.0256	.0756	.1536	.2500	.3456	.4116	.4096	.2916	.1715
	4		.0001	.0016	.0081	.0256	.0625	.1296	.2401	.4096	.6561	.8145
5	0	.7738	.5905	.3277	.1681	.0778	.0313	.0102	.0024	.0003		
	1	.2036	.3281	.4096	.3602	.2592	.1562	.0768	.0284	.0064	.0004	
	2	.0214	.0729	.2048	.3087	.3456	.3125	.2304	.1323	.0512	.0081	.0011
	3	.0011	.0081	.0512	.1323	.2304	.3125	.3456	.3087	.2048	.0729	.0214
	4		.0004	.0064	.0284	.0768	.1562	.2592	.3602	.4096	.3281	.2036
	5			.0003	.0024	.0102	.0313	.0778	.1681	.3277	.5905	.7738
6	0	.7351	.5314	.2621	.1176	.0467	.0156	.0041	.0007	.0001		
	1	.2321	.3543	.3932	.3025	.1866	.0938	.0369	.0102	.0015	.0001	
	2	.0305	.0984	.2458	.3241	.3110	.2344	.1382	.0595	.0154	.0012	.0001
	3	.0021	.0146	.0819	.1852	.2765	.3125	.2765	.1852	.0819	.0146	.0021
	4	.0001	.0012	.0154	.0595	.1382	.2344	.3110	.3241	.2458	.0984	.0305
	5		.0001	.0015	.0102	.0369	.0938	.1866	.3025	.3932	.3543	.2321
	6			.0001	.0007	.0041	.0156	.0467	.1176	.2621	.5314	.7351

n	x											
7	0	.6983	.4783	.2097	.0824	.0280	.0078	.0016	.0002			
	1	.2573	.3720	.3670	.2471	.1306	.0547	.0172	.0036	.0004		
	2	.0406	.1240	.2753	.3177	.2613	.1641	.0774	.0250	.0043	.0002	
	3	.0036	.0230	.1147	.2269	.2903	.2734	.1935	.0972	.0287	.0026	.0002
	4	.0002	.0026	.0287	.0972	.1935	.2734	.2903	.2269	.1147	.0230	.0036
	5		.0002	.0043	.0250	.0774	.1641	.2613	.3177	.2753	.1240	.0406
	6			.0004	.0036	.0172	.0547	.1306	.2471	.3670	.3720	.2573
	7				.0002	.0016	.0078	.0280	.0824	.2097	.4783	.6983
8	0	.6634	.4305	.1678	.0576	.0168	.0039	.0007	.0001			
	1	.2793	.3826	.3355	.1976	.0896	.0312	.0079	.0012	.0001		
	2	.0515	.1488	.2936	.2965	.2090	.1094	.0413	.0100	.0011		
	3	.0054	.0331	.1468	.2541	.2787	.2188	.1239	.0467	.0092	.0004	
	4	.0004	.0046	.0459	.1361	.2322	.2734	.2322	.1361	.0459	.0046	.0004
	5		.0004	.0092	.0467	.1239	.2188	.2787	.2541	.1468	.0331	.0054
	6			.0011	.0100	.0413	.1094	.2090	.2965	.2936	.1488	.0515
	7			.0001	.0012	.0079	.0312	.0896	.1976	.3355	.3826	.2793
	8				.0001	.0007	.0039	.0168	.0576	.1678	.4305	.6634
9	0	.6302	.3874	.1342	.0404	.0101	.0020	.0003				
	1	.2985	.3874	.3020	.1556	.0605	.0176	.0035	.0004			
	2	.0628	.1722	.3020	.2668	.1612	.0703	.0122	.0039	.0003		
	3	.0077	.0446	.1762	.2668	.2508	.1641	.0743	.0210	.0028	.0001	
	4	.0006	.0074	.0661	.1715	.2508	.2461	.1672	.0735	.0165	.0008	.0001
	5		.0008	.0165	.0735	.1672	.2461	.2508	.1715	.0661	.0074	.0004
	6		.0001	.0028	.0210	.0743	.1641	.2508	.2668	.1762	.0446	.0077
	7			.0003	.0039	.0122	.0703	.1612	.2668	.3020	.1722	.0628
	8				.0004	.0035	.0176	.0605	.1556	.3020	.3874	.2985
	9					.0003	.0020	.0101	.0404	.1342	.3874	.6302

Table II. Binomial Probability Values—continued

n	x	.05	.10	.20	.30	.40	.50	.60	.70	.80	.90	.95
10	0	.5987	.3487	.1074	.0282	.0060	.0010	.0001				
	1	.3151	.3874	.2684	.1211	.0403	.0098	.0016	.0001			
	2	.0746	.1937	.3020	.2335	.1209	.0439	.0106	.0014	.0001		
	3	.0105	.0574	.2013	.2668	.2150	.1172	.0425	.0090	.0008		
	4	.0010	.0112	.0881	.2001	.2508	.2051	.1115	.0368	.0055	.0001	
	5	.0001	.0015	.0264	.1029	.2007	.2461	.2007	.1029	.0264	.0015	.0001
	6		.0001	.0055	.0368	.1115	.2051	.2508	.2001	.0881	.0112	.0016
	7			.0008	.0090	.0425	.1172	.2150	.2668	.2013	.0574	.0105
	8			.0001	.0015	.0106	.0439	.1209	.2335	.3020	.1937	.0746
	9				.0001	.0016	.0098	.0403	.1211	.2684	.3874	.3151
	10					.0001	.0010	.0060	.0282	.1074	.3487	.5987
11	0	.5688	.3138	.0859	.0198	.0036	.0005					
	1	.3293	.3835	.2362	.0932	.0266	.0054	.0007				
	2	.0867	.2131	.2953	.1998	.0887	.0269	.0052	.0005			
	3	.0137	.0710	.2215	.2568	.1774	.0806	.0234	.0037	.0002		
	4	.0014	.0158	.1107	.2201	.2365	.1611	.0701	.0173	.0017		
	5	.0001	.0025	.0388	.1321	.2207	.2256	.1471	.0566	.0097	.0003	
	6		.0003	.0097	.0566	.1471	.2256	.2207	.1321	.0388	.0025	.0007
	7			.0017	.0173	.0710	.1611	.2365	.2201	.1107	.0158	.0014
	8			.0002	.0037	.0234	.0806	.1774	.2568	.2215	.0710	.0137
	9				.0005	.0052	.0269	.0887	.1998	.2953	.2131	.0867
	10					.0007	.0054	.0266	.0932	.2362	.3835	.3293
	11						.0005	.0036	.0198	.0859	.3138	.5688
12	0	.5404	.2824	.0687	.0138	.0022	.0002					
	1	.3143	.3766	.2062	.0712	.0174	.0029	.0003				
	2	.0988	.2301	.2835	.1678	.0639	.0161	.0025	.0002			
	3	.0173	.0852	.2362	.2397	.1419	.0537	.0125	.0015	.0001		
	4	.0021	.0213	.1329	.2311	.2128	.1209	.0420	.0078	.0005		
	5	.0002	.0038	.0532	.1585	.2270	.1934	.1009	.0291	.0033		
	6		.0005	.0155	.0792	.1766	.2256	.1766	.0792	.0155	.0005	

n	x	.05	.10	.20	.30	.40	.50	.60	.70	.80	.90	.95
12	7			.0033	.0291	.1009	.1934	.2270	.1585	.0532	.0038	.0002
	8			.0005	.0078	.0420	.1208	.2128	.2311	.1329	.0213	.0021
	9			.0001	.0015	.0125	.0537	.1419	.2397	.2362	.0852	.0173
	10				.0002	.0025	.0161	.0639	.1678	.2835	.2301	.0988
	11					.0003	.0029	.0174	.0712	.2062	.3766	.3413
	12						.0002	.0022	.0138	.0687	.2824	.5404
13	0	.5133	.2542	.0550	.0097	.0013	.0001					
	1	.3512	.3672	.1787	.0540	.0113	.0016	.0001				
	2	.1109	.2448	.2680	.1388	.0453	.0095	.0012	.0001			
	3	.0214	.0997	.2457	.2181	.1107	.0349	.0065	.0006			
	4	.0028	.0277	.1535	.2337	.1845	.0873	.0243	.0034	.0002		
	5	.0003	.0055	.0691	.1803	.2214	.1571	.0656	.0142	.0011		
	6		.0008	.0230	.1030	.1968	.2095	.1312	.0442	.0058	.0001	
	7		.0001	.0058	.0442	.1312	.2095	.1968	.1030	.0230	.0008	
	8			.0011	.0142	.0656	.1571	.2214	.1803	.0691	.0055	.0003
	9			.0002	.0034	.0243	.0873	.1845	.2337	.1535	.0277	.0028
	10				.0006	.0065	.0349	.1107	.2181	.2457	.0997	.0214
	11				.0001	.0012	.0095	.0453	.1388	.2680	.2448	.1109
	12					.0001	.0016	.0113	.0540	.1787	.3672	.3512
	13						.0001	.0013	.0097	.0550	.2542	.5133
14	0	.4877	.2288	.0440	.0068	.0008	.0001					
	1	.3593	.3559	.1539	.0407	.0073	.0009	.0001				
	2	.1229	.2570	.2501	.1134	.0317	.0056	.0006				
	3	.0259	.1142	.2501	.1943	.0845	.0222	.0033	.0002			
	4	.0037	.0349	.1720	.2290	.1549	.0611	.0136	.0014			
	5	.0004	.0078	.0860	.1963	.2066	.1222	.0408	.0066	.0003		
	6		.0013	.0322	.1262	.2066	.1833	.0918	.0232	.0020		
	7		.0002	.0092	.0618	.1574	.2095	.1574	.0618	.0092	.0002	
	8			.0020	.0232	.0918	.1833	.2066	.1262	.0322	.0013	
	9			.0003	.0066	.0408	.1222	.2066	.1963	.0860	.0078	.0004
	10				.0014	.0136	.0611	.1549	.2290	.1720	.0349	.0037
	11				.0002	.0033	.0222	.0845	.1943	.2501	.1142	.0259
	12					.0006	.0056	.0317	.1134	.2501	.2570	.1229
	13					.0001	.0009	.0073	.0407	.1539	.3559	.3593
	14						.0001	.0008	.0068	.0440	.2288	.4877

Table II. Binomial Probability Values—*continued*

n	x	.05	.10	.20	.30	.40	p .50	.60	.70	.80	.90	.95
15	0	.4633	.2059	.0352	.0047	.0005						
	1	.3658	.3432	.1319	.0305	.0047	.0005					
	2	.1348	.2669	.2309	.0916	.0219	.0032	.0003				
	3	.0307	.1285	.2501	.1700	.0634	.0139	.0016	.0001			
	4	.0049	.0428	.1876	.2186	.1268	.0417	.0074	.0006	.0001		
	5	.0006	.0105	.1032	.2061	.1859	.0916	.0245	.0030	.0007		
	6		.0019	.0430	.1472	.2066	.1527	.0612	.0116	.0035	.0003	
	7		.0003	.0138	.0811	.1771	.1964	.1181	.0348	.0138	.0019	
	8			.0035	.0348	.1181	.1964	.1771	.0811	.0430	.0105	
	9			.0007	.0116	.0612	.1527	.2066	.1472	.1032	.0428	
	10			.0001	.0030	.0245	.0916	.1859	.2061	.1876	.1285	.0006
	11				.0006	.0074	.0417	.1268	.2186	.2501	.2669	.0049
	12				.0001	.0016	.0139	.0634	.1700	.2309	.3432	.0307
	13					.0003	.0032	.0219	.0916	.1319	.2059	.1348
	14						.0005	.0047	.0305	.0352		.3658
	15							.0005	.0047			.4633
20	0	.3585	.1216	.0115	.0008							
	1	.3774	.2702	.0576	.0068	.0005						
	2	.1887	.2852	.1369	.0278	.0031	.0002					
	3	.0596	.1901	.2054	.0716	.0124	.0011					
	4	.0133	.0898	.2182	.1304	.0350	.0046	.0003				
	5	.0022	.0319	.1746	.1789	.0746	.0148	.0013				
	6	.0003	.0089	.1091	.1916	.1244	.0370	.0049	.0002			
	7		.0020	.0546	.1643	.1659	.0739	.0146	.0010			
	8		.0004	.0222	.1144	.1797	.1201	.0355	.0039	.0001		
	9		.0001	.0074	.0654	.1597	.1602	.0710	.0120	.0005		
	10			.0020	.0308	.1171	.1762	.1171	.0308	.0020		
	11			.0005	.0120	.0710	.1602	.1597	.0654	.0074	.0001	
	12			.0001	.0039	.0355	.1201	.1797	.1144	.0222	.0004	
	13				.0010	.0146	.0739	.1659	.1643	.0546	.0020	

(continued, $n = 20$)

x											
14				.0002	.0049	.0370	.1244	.1916	.1091	.0089	.0003
15					.0013	.0148	.0746	.1789	.1746	.0319	.0022
16					.0003	.0046	.0350	.1304	.2182	.0898	.0133
17						.0011	.0124	.0716	.2054	.1901	.0596
18						.0002	.0031	.0278	.1369	.2852	.1887
19							.0005	.0068	.0576	.2702	.3774
20								.0008	.0115	.1216	.3585

$n = 25$

x											
0	.2774	.0718	.0038	.0001							
1	.3650	.1994	.0236	.0014							
2	.2305	.2659	.0708	.0074	.0004						
3	.0930	.2265	.1358	.0243	.0019	.0001					
4	.0269	.1384	.1867	.0572	.0071	.0004					
5	.0060	.0646	.1960	.1030	.0199	.0016					
6	.0010	.0239	.1633	.1472	.0442	.0053	.0002				
7	.0001	.0072	.1108	.1712	.0800	.0143	.0009				
8		.0018	.0623	.1651	.1200	.0322	.0031	.0001			
9		.0004	.0294	.1336	.1511	.0609	.0088	.0004			
10		.0001	.0118	.0916	.1612	.0974	.0212	.0013			
11			.0040	.0536	.1465	.1328	.0434	.0042	.0001		
12			.0012	.0268	.1139	.1550	.0760	.0115	.0003		
13			.0003	.0115	.0760	.1550	.1139	.0268	.0012		
14			.0001	.0042	.0434	.1328	.1465	.0536	.0040		
15				.0013	.0212	.0974	.1612	.0916	.0118	.0001	
16				.0004	.0088	.0609	.1511	.1336	.0294	.0004	
17				.0001	.0031	.0322	.1200	.1651	.0623	.0018	
18					.0009	.0143	.0800	.1712	.1108	.0072	.0002
19					.0002	.0053	.0442	.1472	.1633	.0239	.0010
20						.0016	.0199	.1030	.1960	.0646	.0060
21						.0004	.0071	.0572	.1867	.1384	.0269
22						.0001	.0019	.0243	.1358	.2265	.0930
23							.0004	.0074	.0708	.2659	.2305
24								.0014	.0236	.1994	.3650
25								.0001	.0038	.0718	.2774

Table III

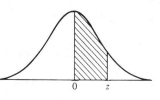

Standardized Normal Distribution—Areas Under
the Standard Normal Curve from 0 to z

z	0	1	2	3	4	5	6	7	8	9
0.0	.0000	.0040	.0080	.0120	.0160	.0199	.0239	.0279	.0319	.0359
0.1	.0398	.0438	.0478	.0517	.0557	.0596	.0636	.0675	.0714	.0754
0.2	.0793	.0832	.0871	.0910	.0948	.0987	.1026	.1064	.1103	.1141
0.3	.1179	.1217	.1255	.1293	.1331	.1368	.1406	.1443	.1480	.1517
0.4	.1554	.1591	.1628	.1664	.1700	.1736	.1772	.1808	.1844	.1879
0.5	.1915	.1950	.1985	.2019	.2054	.2088	.2123	.2157	.2190	.2224
0.6	.2258	.2291	.2324	.2357	.2389	.2422	.2454	.2486	.2518	.2549
0.7	.2580	.2612	.2642	.2673	.2704	.2734	.2764	.2794	.2823	.2852
0.8	.2881	.2910	.2939	.2967	.2966	.3023	.3051	.3078	.3106	.3133
0.9	.3159	.3186	.3212	.3238	.3264	.3289	.3315	.3340	.3365	.3389
1.0	.3413	.3438	.3461	.3485	.3508	.3531	.3554	.3577	.3599	.3621
1.1	.3643	.3665	.3686	.3708	.3729	.3749	.3770	.3790	.3810	.3830
1.2	.3849	.3869	.3888	.3907	.3925	.3944	.3962	.3980	.3997	.4015
1.3	.4032	.4049	.4066	.4082	.4099	.4115	.4131	.4147	.4162	.4177
1.4	.4192	.4207	.4222	.4236	.4251	.4265	.4279	.4292	.4306	.4319
1.5	.4332	.4345	.4357	.4370	.4382	.4394	.4406	.4418	.4429	.4441
1.6	.4452	.4463	.4474	.4484	.4495	.4505	.4515	.4525	.4535	.4545
1.7	.4554	.4564	.4573	.4582	.4591	.4599	.4608	.4616	.4625	.4633
1.8	.4641	.4649	.4656	.4664	.4671	.4678	.4686	.4693	.4699	.4706
1.9	.4713	.4719	.4726	.4732	.4738	.4744	.4750	.4756	.4761	.4767
2.0	.4772	.4778	.4783	.4788	.4793	.4798	.4803	.4808	.4812	.4817
2.1	.4821	.4826	.4830	.4834	.4838	.4842	.4846	.4850	.4854	.4857
2.2	.4861	.4864	.4868	.4871	.4875	.4878	.4881	.4884	.4887	.4890
2.3	.4893	.4896	.4898	.4901	.4904	.4906	.4909	.4911	.4913	.4916
2.4	.4918	.4920	.4922	.4925	.4927	.4949	.4931	.4932	.4934	.4936
2.5	.4938	.4940	.4941	.4943	.4945	.4946	.4948	.4949	.4951	.4952
2.6	.4953	.4955	.4956	.4957	.4959	.4960	.4961	.4962	.4963	.4964
2.7	.4965	.4966	.4967	.4968	.4969	.4970	.4971	.4972	.4973	.4974
2.8	.4974	.4975	.4976	.4977	.4977	.4978	.4979	.4979	.4980	.4981
2.9	.4981	.4982	.4982	.4983	.4984	.4984	.4985	.4985	.4986	.4986
3.0	.4987	.4987	.4987	.4988	.4988	.4989	.4989	.4989	.4990	.4990
3.1	.4990	.4991	.4991	.4991	.4992	.4992	.4992	.4992	.4993	.4993
3.2	.4993	.4993	.4994	.4994	.4994	.4994	.4994	.4995	.4995	.4995
3.3	.4995	.4995	.4995	.4996	.4996	.4996	.4996	.4996	.4996	.4997
3.4	.4997	.4997	.4997	.4997	.4997	.4997	.4997	.4997	.4997	.4998
3.5	.4998	.4998	.4998	.4998	.4998	.4998	.4998	.4998	.4998	.4998
3.6	.4998	.4998	.4999	.4999	.4999	.4999	.4999	.4999	.4999	.4999
3.7	.4999	.4999	.4999	.4999	.4999	.4999	.4999	.4999	.4999	.4999
3.8	.4999	.4999	.4999	.4999	.4999	.4999	.4999	.4999	.4999	.4999
3.9	.5000	.5000	.5000	.5000	.5000	.5000	.5000	.5000	.5000	.5000

Table IV

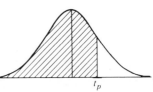

Student's *t*-Distribution—Percentile Values (t_p) with
ν Degrees of Freedom (shaded area = p)

ν	$t_{.995}$	$t_{.99}$	$t_{.975}$	$t_{.95}$	$t_{.90}$	$t_{.80}$	$t_{.75}$	$t_{.70}$	$t_{.60}$	$t_{.55}$
1	63.66	31.82	12.71	6.31	3.08	1.376	1.000	.727	.325	.158
2	9.92	6.96	4.30	2.92	1.89	1.061	.816	.617	.289	.142
3	5.84	4.54	3.18	2.35	1.64	.978	.765	.584	.277	.137
4	4.60	3.75	2.78	2.13	1.53	.941	.741	.569	.271	.134
5	4.03	3.36	2.57	2.02	1.48	.920	.727	.559	.267	.132
6	3.71	3.14	2.45	1.94	1.44	.906	.718	.553	.265	.131
7	3.50	3.00	2.36	1.90	1.42	.896	.711	.549	.263	.130
8	3.36	2.90	2.31	1.86	1.40	.889	.706	.546	.262	.130
9	3.25	2.82	2.26	1.83	1.38	.883	.703	.543	.261	.129
10	3.17	2.76	2.23	1.81	1.37	.879	.700	.542	.260	.129
11	3.11	2.72	2.20	1.80	1.36	.876	.697	.540	.260	.129
12	3.06	2.68	2.18	1.78	1.36	.873	.695	.539	.259	.128
13	3.01	2.65	2.16	1.77	1.35	.870	.694	.538	.259	.128
14	2.98	2.62	2.14	1.76	1.34	.868	.692	.537	.258	.128
15	2.95	2.60	2.13	1.75	1.34	.866	.691	.536	.258	.128
16	2.92	2.58	2.12	1.75	1.34	.865	.690	.535	.258	.128
17	2.90	2.57	2.11	1.74	1.33	.863	.689	.534	.257	.128
18	2.88	2.55	2.10	1.73	1.33	.862	.688	.534	.257	.127
19	2.86	2.54	2.09	1.73	1.33	.861	.688	.533	.257	.127
20	2.84	2.53	2.09	1.72	1.32	.860	.687	.533	.257	.127
21	2.83	2.52	2.08	1.72	1.32	.859	.686	.532	.257	.127
22	2.82	2.51	2.07	1.72	1.32	.858	.686	.532	.256	.127
23	2.81	2.50	2.07	1.71	1.32	.858	.685	.532	.256	.127
24	2.80	2.49	2.06	1.71	1.32	.857	.685	.531	.256	.127
25	2.79	2.48	2.06	1.71	1.32	.856	.684	.531	.256	.127
26	2.78	2.48	2.06	1.71	1.32	.856	.684	.531	.256	.127
27	2.77	2.47	2.05	1.70	1.31	.855	.684	.531	.256	.127
28	2.76	2.47	2.05	1.70	1.31	.855	.683	.530	.256	.127
29	2.76	2.46	2.04	1.70	1.31	.854	.683	.530	.256	.127
30	2.75	2.46	2.04	1.70	1.31	.854	.683	.530	.256	.127
40	2.70	2.42	2.02	1.68	1.30	.851	.681	.529	.255	.126
60	2.66	2.39	2.00	1.67	1.30	.848	.679	.527	.254	.126
120	2.62	2.36	1.98	1.66	1.29	.845	.677	.526	.254	.126
∞	2.58	2.33	1.96	1.645	1.28	.842	.674	.524	.253	.126

Source: Table III of R. A. Fisher and F. Yates: *Statistical Tables for Biological, Agricultural, and Medical Research*, published by Longman Group Ltd., London (previously published by Oliver and Boyd, Edinburgh, Scotland), and by permission of the authors and publishers.

Table V

Chi-Square Distribution—Percentile Values (χ_p^2) with
ν Degrees of Freedom (shaded area $= p$)

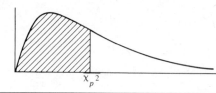

ν	$\chi_{.995}^2$	$\chi_{.99}^2$	$\chi_{.975}^2$	$\chi_{.95}^2$	$\chi_{.90}^2$	$\chi_{.75}^2$	$\chi_{.50}^2$	$\chi_{.25}^2$	$\chi_{.10}^2$	$\chi_{.05}^2$	$\chi_{.025}^2$	$\chi_{.01}^2$	$\chi_{.005}^2$
1	7.88	6.63	5.02	3.84	2.71	1.32	.455	.102	.0158	.0039	.0010	.0002	.0000
2	10.6	9.21	7.38	5.99	4.61	2.77	1.39	.575	.211	.103	.0506	.0201	.0100
3	12.8	11.3	9.35	7.81	6.25	4.11	2.37	1.21	.584	.352	.216	.115	.072
4	14.9	13.3	11.1	9.49	7.78	5.39	3.36	1.92	1.06	.711	.484	.297	.207
5	16.7	15.1	12.8	11.1	9.24	6.63	4.35	2.67	1.61	1.15	.831	.554	.412
6	18.5	16.8	14.4	12.6	10.6	7.84	5.35	3.45	2.20	1.64	1.24	.872	.676
7	20.3	18.5	16.0	14.1	12.0	9.04	6.35	4.25	2.83	2.17	1.69	1.24	.989
8	22.0	20.1	17.5	15.5	13.4	10.2	7.34	5.07	3.49	2.73	2.18	1.65	1.34
9	23.6	21.7	19.0	16.9	14.7	11.4	8.34	5.90	4.17	3.33	2.70	2.09	1.73
10	25.2	23.2	20.5	18.3	16.0	12.5	9.34	6.74	4.87	3.94	3.25	2.56	2.16
11	26.8	24.7	21.9	19.7	17.3	13.7	10.3	7.58	5.58	4.57	3.82	3.05	2.60
12	28.3	26.2	23.3	21.0	18.5	14.8	11.3	8.44	6.30	5.23	4.40	3.57	3.07
13	29.8	27.7	24.7	22.4	19.8	16.0	12.3	9.30	7.04	5.89	5.01	4.11	3.57
14	31.3	29.1	26.1	23.7	21.1	17.1	13.3	10.2	7.79	6.57	5.63	4.66	4.07
15	32.8	30.6	27.5	25.0	22.3	18.2	14.3	11.0	8.55	7.26	6.26	5.23	4.60
16	34.3	32.0	28.8	26.3	23.5	19.4	15.3	11.9	9.31	7.96	6.91	5.81	5.14
17	35.7	33.4	30.2	27.6	24.8	20.5	16.3	12.8	10.1	8.67	7.56	6.41	5.70
18	37.2	34.8	31.5	28.9	26.0	21.6	17.3	13.7	10.9	9.39	8.23	7.01	6.26
19	38.6	36.2	32.9	30.1	27.2	22.7	18.3	14.6	11.7	10.1	8.91	7.63	6.84
20	40.0	37.6	34.2	31.4	28.4	23.8	19.3	15.5	12.4	10.9	9.59	8.26	7.43
21	41.4	38.9	35.5	32.7	29.6	24.9	20.3	16.3	13.2	11.6	10.3	8.90	8.03
22	42.8	40.3	36.8	33.9	30.8	26.0	21.3	17.2	14.0	12.3	11.0	9.54	8.64
23	44.2	41.6	38.1	35.2	32.0	27.1	22.3	18.1	14.8	13.1	11.7	10.2	9.26
24	45.6	43.0	39.4	36.4	33.2	28.2	23.3	19.0	15.7	13.8	12.4	10.9	9.89
25	46.9	44.3	40.6	37.7	34.4	29.3	24.3	19.9	16.5	14.6	13.1	11.5	10.5
26	48.3	45.6	41.9	38.9	35.6	30.4	25.3	20.8	17.3	15.4	13.8	12.2	11.2
27	49.6	47.0	43.2	40.1	36.7	31.5	26.3	21.7	18.1	16.2	14.6	12.9	11.8
28	51.0	48.3	44.5	41.3	37.9	32.6	27.3	22.7	18.9	16.9	15.3	13.6	12.5
29	52.3	49.6	45.7	42.6	39.1	33.7	28.3	23.6	19.8	17.7	16.0	14.3	13.1
30	53.7	50.9	47.0	43.8	40.3	34.8	29.3	24.5	20.6	18.5	16.8	15.0	13.8
40	66.8	63.7	59.3	55.8	51.8	45.6	39.3	33.7	29.1	26.5	24.4	22.2	20.7
50	79.5	76.2	71.4	67.5	63.2	56.3	49.3	42.9	37.7	34.8	32.4	29.7	28.0
60	92.0	88.4	83.3	79.1	74.4	67.0	59.3	52.3	46.5	43.2	40.5	37.5	35.5
70	104.2	100.4	95.0	90.5	85.5	77.6	69.3	61.7	55.3	51.7	48.8	45.4	43.3
80	116.3	112.3	106.6	101.9	96.6	88.1	79.3	71.1	64.3	60.4	57.2	53.5	51.2
90	128.3	124.1	118.1	113.1	107.6	98.6	89.3	80.6	73.3	69.1	65.6	61.8	59.2
100	140.2	135.8	129.6	124.3	118.5	109.1	99.3	90.1	82.4	77.9	74.2	70.1	67.3

Table VI. Mann–Whitney U-Test
(One-tailed test at .05 level; two-tailed test at .10 level)

| | | | | | | | | | | | | | n | | | | | | | |
m	1	2	3	4	5	6	7	8	9	10	11	12	13	14	15	16	17	18	19	20
1	—																			
2	—	—																		
3	—	—	0																	
4	—	—	0	1																
5	—	0	1	2	4															
6	—	0	2	3	5	7														
7	—	0	2	4	6	8	11													
8	—	1	3	5	8	10	13	15												
9	—	1	4	6	9	12	15	18	21											
10	—	1	4	7	11	14	17	20	24	27										
11	—	1	5	8	12	16	19	23	27	31	34									
12	—	2	5	9	13	17	21	26	30	34	38	42								
13	—	2	6	10	15	19	24	28	33	37	42	47	51							
14	—	3	7	11	16	21	26	31	36	41	46	51	56	61						
15	—	3	7	12	18	23	28	33	39	44	50	55	61	66	72					
16	—	3	8	14	19	25	30	36	42	48	54	60	65	71	77	83				
17	—	3	9	15	20	26	33	39	45	51	57	64	70	77	83	89	96			
18	—	4	9	16	22	28	35	41	48	55	61	68	75	82	88	95	102	109		
19	0	4	10	17	23	30	37	44	51	58	65	72	80	87	94	101	109	116	123	
20	0	4	11	18	25	32	39	47	54	62	69	77	84	92	100	107	115	123	130	138
21	0	5	11	19	26	34	41	49	57	65	73	81	89	97	105	113	121	130	138	146
22	0	5	12	20	28	36	44	52	60	68	77	85	94	102	111	119	128	136	145	154
23	0	5	13	21	29	37	46	54	63	72	81	90	98	107	116	125	134	143	152	161
24	0	6	13	22	30	39	48	57	66	75	85	94	103	113	122	131	141	150	160	162
25	0	6	14	23	32	41	50	60	69	79	89	98	108	118	128	137	147	157	167	177
26	0	6	15	24	33	43	53	62	72	82	92	103	113	123	133	143	154	164	174	185
27	0	7	15	25	35	45	55	65	75	86	96	107	117	128	139	149	160	171	182	192
28	0	7	16	26	36	46	57	68	78	89	100	111	122	133	144	156	167	178	180	200
29	0	7	17	27	38	48	59	70	82	93	104	116	127	138	150	162	173	185	196	208
30	0	7	17	28	39	50	61	73	85	96	108	120	132	144	156	168	180	192	204	216
31	0	8	18	29	40	52	64	76	88	100	112	124	136	149	161	174	186	199	211	224
32	0	8	19	30	42	54	66	78	91	103	116	128	141	154	167	180	193	206	218	231
33	0	8	19	31	43	56	68	81	94	107	120	133	146	159	172	186	199	212	226	239
34	0	9	20	32	45	57	70	84	97	110	124	137	151	164	178	192	206	219	233	247
35	0	9	21	33	46	59	73	86	100	114	128	141	156	170	184	198	212	226	241	255
36	0	9	21	34	48	61	75	89	103	117	131	146	160	175	189	204	219	233	248	263
37	0	10	22	35	49	63	77	91	106	121	135	150	165	180	195	210	225	240	255	271
38	0	10	23	36	50	65	79	94	109	124	139	154	170	185	201	216	232	247	263	278
39	1	10	23	38	52	67	82	97	112	128	143	159	175	190	206	222	238	254	270	286
40	1	11	24	39	53	68	84	99	115	131	147	163	179	196	212	228	245	261	278	294

Source: American Statistical Association Journal (September 1964), pp. 927–932.

Table VI (*cont.*) (One-tailed test at .025 level; two-tailed test at .05 level)

m	1	2	3	4	5	6	7	8	9	10	11	12	13	14	15	16	17	18	19	20
1	—																			
2	—	—																		
3	—	—	—																	
4	—	—	—	0																
5	—	—	0	1	2															
6	—	—	1	2	3	5														
7	—	—	1	3	5	6	8													
8	—	0	2	4	6	8	10	13												
9	—	0	2	4	7	10	12	15	17											
10	—	0	3	5	8	11	14	17	20	23										
11	—	0	3	6	9	13	16	19	23	26	30									
12	—	1	4	7	11	14	18	22	26	29	33	37								
13	—	1	4	8	12	16	20	24	28	33	37	41	45							
14	—	1	5	9	13	17	22	26	31	36	40	45	50	55						
15	—	1	5	10	14	19	24	29	34	39	44	49	54	59	64					
16	—	1	6	11	15	21	26	31	37	42	47	53	59	64	70	75				
17	—	2	6	11	17	22	28	34	39	45	51	57	63	69	75	81	87			
18	—	2	7	12	18	24	30	36	42	48	55	61	67	74	80	86	93	99		
19	—	2	7	13	19	25	32	38	45	52	58	65	72	78	85	92	99	106	113	
20	—	2	8	14	20	27	34	41	48	55	62	69	76	83	90	98	105	112	119	127
21	—	3	8	15	22	29	36	43	50	58	65	73	80	88	96	103	111	119	126	134
22	—	3	9	16	23	30	38	45	53	61	69	77	85	93	101	109	117	125	133	141
23	–	3	9	17	24	32	40	48	56	64	73	81	89	98	106	115	123	132	140	149
24	—	3	10	17	25	33	42	50	59	67	76	85	94	102	111	120	129	138	147	156
25	—	3	10	18	27	35	44	53	62	71	80	89	98	107	117	126	135	145	154	163
26	—	4	11	19	28	37	46	55	64	74	83	93	102	112	122	132	141	151	161	171
27	—	4	11	20	29	38	48	57	67	77	87	97	107	117	127	137	147	158	168	178
28	—	4	12	21	30	40	50	60	70	80	90	101	111	122	132	143	154	164	175	186
29	—	4	13	22	32	42	52	62	73	83	94	105	116	127	138	149	160	171	182	193
30	—	5	13	23	33	43	54	65	76	87	98	109	120	131	143	154	166	177	189	200
31	—	5	14	24	34	45	56	67	78	90	101	113	125	136	148	160	172	184	196	208
32	—	5	14	24	35	46	58	69	81	93	105	117	129	141	153	166	178	190	203	215
33	—	5	15	25	37	48	60	72	84	96	108	121	133	146	159	171	184	197	210	222
34	—	5	15	26	38	50	62	74	87	99	112	125	138	151	164	177	190	203	217	230
35	—	6	16	27	39	51	64	77	89	103	116	129	142	156	169	183	196	210	224	237
36	—	6	16	28	40	53	66	79	92	106	119	133	147	161	174	188	202	216	231	245
37	—	6	17	29	41	55	68	81	95	109	123	137	151	165	180	194	209	223	238	252
38	—	6	17	30	43	56	70	84	98	112	127	141	156	170	185	200	215	230	245	259
39	0	7	18	31	44	58	72	86	101	115	130	145	160	175	190	206	221	236	252	267
40	0	7	18	31	45	59	74	89	103	119	134	149	165	180	196	211	227	243	258	274

Table VII. Random Numbers

10097	32533	76520	13586	34673	54876	80959	09117	39292	74945
37542	04805	64894	74296	24805	24037	20636	10402	00822	91665
08422	68953	19645	09303	23209	02560	15953	34764	35080	33606
99019	02529	09376	70715	38311	31165	88676	74397	04436	27659
12807	99970	80157	36147	64032	36653	98951	16877	12171	76833
66065	74717	34072	76850	36697	36170	65813	39885	11199	29170
31060	10805	45571	82406	35303	42614	86799	07439	23403	09732
85269	77602	02051	65692	68665	74818	73053	85247	18623	88579
63573	32135	05325	47048	90553	57548	28468	28709	83491	25624
73796	45753	03529	64778	35808	34282	60935	20344	35273	88435
98520	17767	14905	68607	22109	40558	60970	93433	50500	73998
11805	05431	39808	27732	50725	68248	29405	24201	52775	67851
83452	99634	06288	98083	13746	70078	18475	40610	68711	77817
88685	40200	86507	58401	36766	67951	90364	76493	29609	11062
99594	67348	87517	64969	91826	08928	93785	61368	23478	34113
65481	17674	17468	50950	58047	76974	73039	57186	40218	16544
80124	35635	17727	08015	45318	22374	21115	78253	14385	53763
74350	99817	77402	77214	43236	00210	45521	64237	96286	02655
69916	26803	66252	29148	36936	87203	76621	13990	94400	56418
09893	20505	14225	68514	46427	56788	96297	78822	54382	14598
91499	14523	68479	27686	46162	83554	94750	89923	37089	20048
80336	94598	26940	36858	70297	34135	53140	33340	42050	82341
44104	81949	85157	47954	32979	26575	57600	40881	22222	06413
12550	73742	11100	02040	12860	74697	96644	89439	28707	25815
63606	49329	16505	34484	40219	52563	43651	77082	07207	31790
61196	90446	26457	47774	51924	33729	65394	59593	42582	60527
15474	45266	95270	79953	59367	83848	82396	10118	33211	59466
94557	28573	67897	54387	54622	44431	91190	42592	92927	45973
42481	16213	97344	08721	16868	48767	03071	12059	25701	46670
23523	78317	73208	89837	68935	91416	26252	29663	05522	82562
04493	52494	75246	33824	45862	51025	61962	79335	65337	12472
00549	97654	64051	88159	96119	63896	54692	82391	23287	29529
35963	15307	26898	09354	33351	35462	77974	50024	90103	39333
59808	08391	45427	26842	83609	49700	13021	24892	78565	20106
46058	85236	01390	92286	77281	44077	93910	83647	70617	42941
32179	00597	87379	25241	05567	07007	86743	17157	85394	11838
69234	61406	20117	45204	15956	60000	18743	92423	97118	96338
19565	41430	01758	75379	40419	21585	66674	36806	84962	85207
45155	14938	19476	07246	43667	94543	59047	90033	20826	69541
94864	31994	36168	10851	34888	81553	01540	35456	05014	51176
98086	24826	45240	28404	44999	08896	39094	73407	35441	31880
33185	16232	41941	50949	89435	48581	88695	41994	37548	73043
80951	00406	96382	70774	20151	23387	25016	25298	94624	61171
79752	49140	71961	28296	69861	02591	74852	20539	00387	59579
18633	32537	98145	06571	31010	24674	05455	61427	77938	91936
74029	43902	77557	32270	97790	17119	52527	58021	80814	51748
54178	45611	80993	37143	05335	12969	56127	19255	36040	90324
11664	49883	52079	84879	59381	71539	09973	33440	88461	23356
48324	77928	31249	64710	02295	36870	32307	57546	15020	09994
69074	94138	87637	91976	35584	04401	10518	21615	01848	76938

Source: This table is reproduced with permission from the RAND Corporation: *A Million Random Digits*, 1955.

Answers to Selected Problems

2.3.1 (a) Discrete (b) No. Inasmuch as most of the 30 values are different, an ungrouped frequency distribution would not give us much more information.

(c)

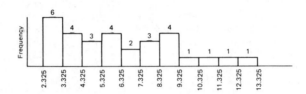

2.7.1 .0611; .0476; .2075; .6837
2.7.3 (c) $1 - \frac{26}{30} = \frac{4}{30}$ (e) $\frac{15}{30}$
2.7.7 (a) Skewed to the right (b) Approximately normal
3.3.2 15.5–19.5; 22.73
3.5.1 (a) 5 (b) 4.5 (c) 4.1 (f) 4.5–5.5 (h) 4.1
3.5.2 23.08
3.7.1 76.38($P_{.75}$)
3.8.3 (a) 4.05 (b) 60%
3.8.4 29; 7.8
3.8.7 (a) 100%; 92.7%
3.9.1 159.6
3.10.1 (a) 1.94; 19.4 (c) (120.8, 198.4)
3.10.2 .7
3.13.1 6.46; 4.5; 4
3.13.3 17.57; 14.25; 20.81
3.13.4 (a) 29.175 (b) 26.925
3.13.5 (a) 17.52 (b) 35.0496 (c) 5.92 (d) 75.36
3.13.6 84
3.13.8 (a) 1,026.42; 8.33 (b) 71%
3.13.9 (a) You need to know that $\bar{x} = 81$ and $s = 12.81$ (c) 67% (d) 67%
3.13.10 $w_i = 2.5x_i + 25$
4.3.2 (a)

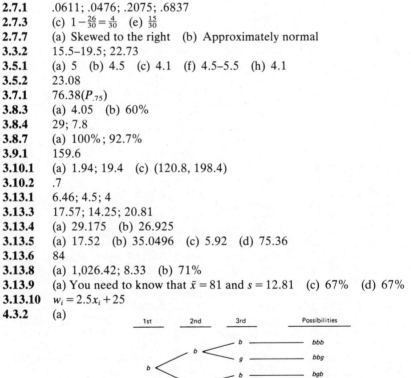

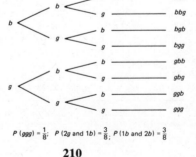

210

4.3.3   No

4.3.4   (a) $\frac{6}{11}$ (b) $\frac{5}{11}$ (d) $\frac{5}{11}$ (e) $\frac{8}{11}$ (f) $\frac{8}{11}$

4.3.5   82%

4.3.6   Uniform

4.4.1 $4 \times 3 \times 2 \times 1 = 24$

4.4.2 $4 \times 4 \times 4 - 1 = 63$

4.4.3   (a) $(.015)(.015)$ (b) $1 - (.015)(.015)$

4.4.4 $\frac{1}{4}$

4.4.5   (a) .1035   (b) .77

4.5.2   210; 210

4.6.1   (a) $\frac{1}{4}$ (b) $\frac{3}{8}$ (c) $\frac{1}{4}$ (e) $\frac{1}{16}$

4.6.2   (a) $\frac{1}{16}$ (c) $\frac{15}{16}$

4.6.4 $(.10)^3; \ 3(.10)^2(.90)^1; \ 1 - (.90)^3; \ 1 - (.90)^3 + 3(.10)^1(.90)^2$

4.7.1   1.50; 2.70

4.7.3   3

4.7.7   (a) 7   (c) $\frac{20}{36}; \frac{6}{36}; \frac{14}{36}; \frac{18}{36}$

4.10.4   (a) .60   (b) .08   (c) .032   (d) .7229

4.10.5   (a) $\frac{41}{87}$ (b) $\frac{19}{87}$ (c) $\frac{12}{87}$

4.10.6   (a) $\frac{1}{50}$ (b) $\frac{4}{50}$ (c) $\frac{13}{50}$

4.10.8   (a) 56

4.10.9   (a) $(.25)^4 = .003962$ (e) .316406

4.10.10   (c) 1   (d) .7383

4.10.11   3.2

4.10.12   (a) $-.0526$

5.2.2   (b) $\binom{10}{3}(.50)^3(.50)^7 = .117$ (d) $\binom{10}{10}(.50)^{10}(.50)^0 = .001$ (e) 5

5.2.4   (a) 5; 1.58
            (b) $P(3.42 \le x \le 6.58) = P(x = 4) + P(x = 5) + P(x = 6) = .656$

5.3.2   It must also be rejected. A one-sided test to the right means that the alternative hypothesis was of the form "$P >$"; in this case, $P > .60$. Since the null hypothesis was rejected, $P = .50$ is also rejected.

5.3.5   (a) Retain   (b) Reject

5.3.9   .471; .811

5.3.10   .005; .344

5.4.1   (a) .037   (b) (.083, .157)   (d) 1,014

5.4.2   (a) (.17, .82)   (c) (.47, .53)

5.4.3   Critical region $= \{0–11, \ 25–30\}$

5.4.5 $\bar{p} \ge .60$; 60 or more

5.7.1   .088

5.7.2   .121

5.7.4 $(.11, 3.89) \Rightarrow 77.1$ per cent

5.7.5   167

5.7.6   (a) No   (b) No   (c) No

5.7.7   (a) No   (b) Yes   (c) No

5.7.8   (a) $P = .60$ (b) Critical region $= \{0, \ 1, \ 2, \ 10\}$; $\beta_{.50} = .944$; $\beta_{.30} = .617$ (c) $P = .60$; critical region $= \{0–9, \ 21–25\}$; $\beta_{.50} = .884$; $\beta_{.30} = .189$

5.7.10   (a) $.55 < p < .77$ using Figure 5.4.2; $.572 < p < .768$ as $.67 \pm 1.96(.05)$ is the interval about p (b) 2,211

5.8.1   (a) Mean   (b) Uniform distribution of sales

5.8.3   (a) $\mu_x = 300$; $\sigma_x = 70.71$;   range $= 200$ (b) $y_i = \dfrac{x_i}{100} - 3$; $\mu_y = 0$; $\sigma_y =$ .7071

5.8.4   87; 57

5.8.5   60

5.8.6   (a) .30   (b) .025   (c) .975

5.8.7   (a) .000   (b) .000   (c) .029   (d) .000

5.8.8   .911

5.8.9   Cannot reject for (a) or (b); .793 for both (a) and (b)—did you use the rule for complementary events or did you do a lot of adding?

5.8.10 $.24 < p < .36$; $.22 < p < .37$

5.8.11   (120, 200)

6.1.1   (a) (46, 94)   (b) 34.15%

6.1.2   Neither

6.2.1   (a) .977   (d) 478   (f) ?   (h) 225; 83.32%

6.2.2   Approximately 27 years

6.2.3   (a) .2960   (b) 13.51%   (c) 233.34

6.3.1   (a) .5000   (b) Approximately .2810   (c) .3478

6.3.2   (a) 1.33   (b) 150   (d) Approximately 205   (e) .0038

6.3.3   (a) .249; .2482   (b) .244; .2482   (c) .705; .6970

6.3.4   83

6.4.1   Yes

6.4.4   Retain

6.4.5   Reject

6.4.6   (a) Reject   (c) Retain

6.4.7   (a) .9773   (c) .3108

6.5.1   (a) (52.23, 59.77)

6.5.2   Point estimate (she was not)

6.5.4   126

6.6.3   (a) $201.45 < \mu < 218.55$; $196 < \mu < 224$

6.9.1   (a) 3   (c) 2

6.9.2   (1,500, 1,700) contains 68.3% of the total area

6.9.3   34.13%; 47.72%

6.9.4   (a) 0   (d) 3

6.9.6   (a) .609   (c) .409

6.9.7   .9544

6.9.8 $\alpha \le .05 \Rightarrow$ critical region $= \{0\text{–}24, 40\text{–}60\}$, and acceptance region $= \{25\text{–}39\}$

6.9.10   190

6.9.11   Retain; reject

6.9.14   Yes

6.9.15   (a) 2   (b) Not given; $9.32 < \mu < 11.0$ (c) $46,600 < \text{contributions} < 55,400$

6.9.16   Reject

6.9.17   Retain ($t = -1.50$)

6.9.18   (17.46, 32.54)

7.2.1　Retain
7.2.2　Retain
7.2.3　No
7.3.1　Yes
7.3.2　Independent
7.3.4　No
7.4.1　Retain
7.5.1　Significant
7.8.1　No
7.8.2　No
7.8.3　Yes
7.8.4　Retain
7.8.5　No $(6.4 > 5.99)$
7.8.7　(a) $H_A : P(\text{male more aggressive}) > .50$; retain H_0 as 6 does not fall in the critical region $\{8, 9, 10\}$　(b) Retain
7.8.8　No

Index